Vulture Culture

Studies in the
Postmodern Theory of Education

Joe L. Kincheloe and Shirley R. Steinberg
General Editors

Vol. 152

PETER LANG
New York • Washington, D.C./Baltimore • Bern
Frankfurt am Main • Berlin • Brussels • Vienna • Oxford

Christine M. Quail, Kathalene A. Razzano, and Loubna H. Skalli

Vulture Culture

The Politics and Pedagogy of Daytime Television Talk Shows

FOREWORD BY
Rhonda Hammer and Douglas Kellner

PETER LANG
New York • Washington, D.C./Baltimore • Bern
Frankfurt am Main • Berlin • Brussels • Vienna • Oxford

Library of Congress Cataloging-in-Publication Data
Quail, Christine M.
Vulture culture: the politics and pedagogy of daytime television talk shows /
Christine M. Quail, Kathalene A. Razzano, Loubna H. Skalli.
p. cm. — (Counterpoints)
Includes bibliographical references and index.
1. Talk shows—United States. 2. Talk shows—Social aspects—United States.
I. Razzano, Kathalene A. II. Skalli, Loubna H.
III. Title. IV. Series: Counterpoints (New York, N.Y.)
PN1992.8.T3Q35   791.45'6—dc22   2005022534
ISBN 0-8204-5011-1
ISSN 1058-1634

Bibliographic information published by **Die Deutsche Bibliothek**.
Die Deutsche Bibliothek lists this publication in the "Deutsche
Nationalbibliografie"; detailed bibliographic data is available
on the Internet at http://dnb.ddb.de/.

Cover design by Joni Holst

To Ian, Luna, Elaine & Joe
—C.Q.

To Dan, Christy, my family, friends, and the students and faculty of the
Cultural Studies Program at George Mason University
—K.R.

To my Father, John, A & Z
—L.S.

CONTENTS

FOREWORD

BY

RHONDA HAMMER AND DOUGLAS KELLNER

C ritical writers, like bell hooks, have described the con–
temporary United States as a "hedonistic consumer culture"
that is inscribed by an escalating politics of greed (2000, p.
63). Within the terrain of media culture, hedonism, greed, and social
pathology are evident everyday on mainstream television. TV is a
major source of pedagogy and socialization, yet, all too frequently,
studies of the hegemonic authority of media culture, especially
television, are over-simplified and/or dismissed. Indeed, the complex
and multidimensional nature of media culture, especially those
which for critics comprise the so-called "low" forms, have often
evaded critical investigation. This omission is due, in large part,
to the characterization of such media as "trash," a categorization
that has been etched into the public consciousness—often with the
complicity of media institutions themselves—and which has largely
protected them from the kinds of serious scrutiny they deserve.

Moreover, it is interesting, but hardly surprising, to note that
much of what has been relegated to the lower rungs of media studies are those that are generally regarded as being women's domains,
which encompass much of daytime television. Although the influence, audience numbers, and profit margins often exceed many other television genres, it appears that a great deal of academic media
criticism is reserved for less blatantly feminized gendered forms and
media phenomena. Furthermore, although there has been a growing

body of literature that critically examines daytime television, particularly that of the soap opera, except for mostly uncritical journalistic scans and empirical content or audience analysis, there is little on the phenomenon of daytime talk television, a genre that has proliferated over the years.

Christine Quail, Kathalene Razzano, and Loubna Skalli in their engaging text *Vulture Culture* have provided an important contribution to both the fields of critical media studies and the analysis of contemporary talk television. Their employment of discursive arguments draws and expands upon a wide-ranging field of transdisciplinary theories, which include political economy, cultural studies, critical pedagogy, Marxism, and feminisms as well as the philosophical notions of "power and discipline" of Michel Foucault, Guy Debord's classic conceptualizations of "the society of the spectacle," and the insights of Mikhail Bakhtin on the role of the carnival in participatory culture. Their effective practical application of such complex theories to a seemingly mundane domain reveals the polysemic and paradoxical nature of popular culture, as well as demonstrating the relevance of critical theory to everyday realities. The authors' development of this kind of multiperspectival approach within the context of what they describe as "vulture culture" induces us to read such bastions of daytime talk TV as *Oprah*, *Dr. Phil*, and other popular shows in a new light.

In addition to drawing on discourse theory and cultural studies, the authors show how political economy is crucial to understanding talk television. Production and distribution of the programs are controlled by an ever-shrinking number of corporations, reflecting the concentration of ownership in all forms of media. The commodity form of programming for profit and selling shows to advertisers shapes increasingly commercialization of the programs and the culture, reproducing the hegemony of corporate capitalism. The shows are profit-machines which are cheap to produce, using non-paid ordinary people or celebrities seeking publicity, and they are often tied to commodities like Oprah's book club or Dr. Phil's numerous book scams.

The need for critical analysis of talk television seems especially warranted given the recognition of its significance and ideological power by presidential candidates of both parties. Although Bill Clinton broke with traditions by appearing on the *Arsenio* night-time talk show and MTV in 1992, it was not until 2000 that the influence of daytime advice talk television programming was officially acknowledged through appearances by George W. Bush and Al Gore on the *Oprah* show, and both Bush and John Kerry appeared on the *Dr. Phil* show in the 2004 campaign. Yet, as the authors point out, in 1995 a conservative coalition Empower America, comprised of both Republicans and Democrats like William Bennett and Joe Lieberman, condemned talk shows for promoting "cultural rot" (49). Since then, there has been a decline of the "trash talk" television shows like *Jerry Springer* and increase of advice shows like *Dr. Phil*.

The notion of "vulture culture" as a complex set of ideological practices "by which media scavenges the narratives, discourses, knowledges and everyday common-sense of our culture and presents them back as information, spectacle and entertainment," provides the pattern which connects the authors' diverse series of arguments related to different dimensions of talk show television. The term "vulture culture" seems to express the manner in which the complex and often contradictory nature of globalized, consumer, capitalist values and beliefs are reproduced in both the form and content of popular daytime television as natural and "commonsense."

Talk TV is parasitic on social problems and misery caused in large part by social inequalities and the damage of poverty and lack of education. Even so, the major programs dedicated to advice and everyday life target individual failings and offer largely individual solutions to a wide range of problems, solutions that reproduce dominant ideology and forms of thought and behavior.

Moreover, on daytime talk television, the majority of the guests are women and girls or feminized men, while the host and experts, regardless of gender, embody and uphold traditional patriarchal and dominant middle-class codes. The class bias makes working-class people feel inferior and sets up middle-class and professional people as the social norm and ideal. Importantly, as the authors argue, the politics of difference, especially in relation to class, race, gender, and sexuality are effectively obscured and depicted as one-dimensional, psychological, personal problems. This approach tends to blame the victim rather than critique the socio-political and economic contexts that mediate these kinds of pathologies. Moreover, the constellations of aberrant social types and behaviors that are the topic of many of the shows reify the demonization of marginalized groups. In particular, the text demonstrates how single mothers (predominantly the working poor and lower classes) and youths (especially, teenage girls) are favorite targets of daytime talk television. Discussions of genres, which the authors identify as teens-out-of-control [TOOC], and the escalating numbers of paternity themed shows, also tend to reinforce dominant, conservative traditional family values that maintain stereotypical gendered relations. Hence, absurd and impossible imaginary standards of idealized images of fathers and mothers—and rigid, bifurcated notions of masculinity and femininity—are further reified. Racist and heterosexist assumptions are often inferentially if not overtly reproduced in depictions of heterosexual families as "normal" and gay sexuality as deviant, while extremely negative depictions of people of color and underclass people multiply.

Class, race, gender, and hetrosexualist bias, however, is often subtly communicated in these shows, masked by an ideology of democratic populism that displays a multicultural rainbow of diversity, often with hosts of color like Oprah Winfrey or Montel Williams. These hosts tend to reinforce the American myth that anyone can pull themselves up by their "bootstraps" and can overcome racial (class, gender, or sexual) inequalities through individual attitude, perseverance and moral character (Jhally and Lewis, 1992).

Moreover, as the authors point out, individual authority figures often in the guise of celebrity hosts or guests, as well as slews of so-called professional experts, legitimate the ideologies of individualism and the naturalization of elite hegemonic power, which negates inquiries into social and public responsibilities for transforming social conditions to alleviate oppression and suffering. In this sense, talk TV, as a form of infotainment (i.e., information blended with entertainment) serves as an expansive advertisement for not only its sponsors, but also for the commercial products which it incessantly hypes, as well as the books and services of the hosts and so-called experts, and the commoditization of the viewers themselves who are delivered to sponsors through their TV-watching activity.

Hence, talk television as media spectacle is itself a valuable commodity for the multinational corporations which own and produce them and the laissez-faire and individualistic capitalist values the shows espouse. Media spectacles mesmerize audiences with the sensationalistic news of the day (the O.J. Simpson trial, the Clinton sex scandals, the celebrity trials of the moment, and the spectacles of sports and entertainment which dominate everyday life in consumer and corporate capitalism (Kellner, 2003). The real material conditions of the relationships between poverty, rising unemployment, out-sourcing of jobs, the decimation of social assistance and education programs, and the social conditions of escalating violence have no place in the narcissistic celebrity obsessed domain of the talk television spectacle.

Indeed, celebrities are the icons of media culture, the gods and goddesses of everyday life, and ordinary people are positioned as the worshippers of these celebrities and pawns of "experts" who tell them how to solve their problems and live their lives. In this sense, the popularity of daytime talk television serves as a mode of distraction, in that it encourages a politics of individualistic guilt, envy, and ameliorative action. Rather than teaching audiences how to think critically about the power relations which structure their world and the social conditions which help produce their problems, audiences are taught to focus on their own weaknesses and vulnerabilities and taught how to conform to social norms and dominant modes of thought and behavior.

The pedagogy of talk TV is conformist and reproduces existing relations of power and domination. Although many studies of television focus on the programs as sites of pleasure or as a democratic public sphere, the authors of *Vulture Culture* espouse a dialectical approach that examines the manner in which daytime talk television is both compelling and repellent. Although talk TV promises to provide a democratic space for public debate, it often exploits its marginalized guests and presents them as abnormal and as freaks, at odds with the so-called normalized experiences and values of the hosts, experts, and audience members. This depiction, the authors argue, is another example of how vulture culture operates in daytime television and manages to maintain dominant ideologies of power and control.

Of course, there are positive moments of talk television. Pioneering talk show host Phil Donohue initiated a liberal mode of

TV discussion shows focusing on individual and social problems in 1968, and Oprah Winfrey has probably been the most successful and influential TV talk host and personality in history. These shows discussed issues often neglected by mainstream media and promoted thought and dialogue on many important issues. The more carni-valesque "trash TV" of the *Jerry Springer* variety that mushroomed in the 1990s had transgressive moments, gave voice to individuals and issues often suppressed by mainstream culture, and dramatically presented the problem of male violence against women and family terrorism usually neglected by mainstream culture (Hammer, 2002). The advice shows of even so crass and exploitative a host as *Dr. Phil* provides useful information, as his January 2005 series on weight reduction dramatizes the problems of obesity and the need to deal with the problem. Even so, in addressing this problem, he shame-lessly hawks his own book, TV show, and Website, and thus himself as the solution.

Furthermore, the conformist pedagogy usually preached on talk TV, the imposing of experts and audiences and submission of help-less people to societal authority figures, and relentless mainstreaming of middle-class and commercial values, render the talk shows ulti-mately a means of social control and normalization, as the authors of *Vulture Culture* argue. The following studies engage an issue of great importance and should help promote critical thought and reflection on important forms of pedagogy often overlooked and some of the problems with a familiar form of media culture.

REFERENCES

Hammer, Rhonda. (2002). *Antifeminism and Family Terrorism: A Critical Feminist Perspective.* Lanham, MD: Rowman and Littlefield.

hooks, bell. (2000). *Where We Stand: Class Matters.* New York and London: Routledge.

Jhally, Sut and Justin Lewis. (1992). *Enlightened Racism: The Bill Cosby show, Audiences, and the Myth of the American Dream.* Boulder: Westview.

Kellner, Douglas. (2003). *Media Spectacle.* New York and London: Routledge.

1

Vulture Culture:
An Introduction

V*ulture Culture* is a project that has brought the three authors of this book together in an attempt to understand the wide appeal of the programs that everybody loves to hate: daytime television talk shows. To be fair, it was not the book project that initially brought us together. As fellow researchers, we had already recognized common intellectual pursuits and theoretical affinities in our approaches to media studies. However, it was the insights of Professors Joe L. Kincheloe and Shirley R. Steinberg that finally gave this project form and shape. In working with each of us, and overseeing our individual research, they encouraged us to extend our independent efforts into a collaborative enterprise around the talk show—and we thank them.

The questions we were originally asking focused primarily on the shows themselves, on their production, reception, and content as they intersect with issues of race, class, gender, and sexual orientation. Over the five years working on this project, we became increasingly aware that popular daytime shows cannot be understood in isolation as a common television form.

Unquestionably, these programs are ubiquitous, popular, and pervasive components of television programming that achieved an uncanny level of popularity in contemporary culture. The shows have become an intricate part of a larger media landscape spawning Jerry Springer's *Ringmaster*, Oprah's Book Club, Rosie Dolls,

Ellen e-Bay auctions, Dr. Phil McGraw's *Ultimate Weight Solution*, and many other commercial commodities. However, daytime talk shows present us with something of a puzzle: talk shows are but components, although central, in a much larger media culture of (re)production, circulation, and consumption of various types of knowledge.

We realized that we were dealing with a culture that was much larger and more complex than generally assumed: it is a culture with its own logic, topicality, values, market, and audiences. To fully understand this culture, we need to develop a critical pedagogical approach. This approach begins with the recognition that vulture culture sustains itself by constantly reinventing itself. Season after season, these programs participate in as well as shape, inform, and are informed by debates around issues and concerns of the day—both national (political) and personal (as if we can separate the two!). These include everything from obsession with appearances to national debates on family, including child welfare and paternity. The shows often present themselves in a way that surprises their viewers, delights their fans, and increases the shows' material gains. This surprise, delight and, perhaps most importantly, material gain have accompanied the dramatic debut of the 2004–2005 daytime talk show season.

On September 13, 2004, Oprah Winfrey started her nineteenth season with one of the most spectacular events yet on the daytime talk show. Winfrey surprised 267 members of her studio audience with keys to brand-new sport sedans—courtesy of Pontiac. If we are left to wonder why Pontiac would part with 267 vehicles worth $7 million, we need only contemplate the power of Winfrey's celebrity. According to Jim Holbrook, president of The Zipatoni Co., a marketing agency, "Oprah is a celebrity with instant credibility. . .The power of this is incredible, and incredibly hard to replicate."[1] This was an event that celebrates a form of alliance between media celebrities, television networks, advertising, and consumer industries. It was an "event that marketing executives say could set a new bar for product placement."[2]

Almost a week later, Dr. Phil McGraw addressed American viewers about family values and parenthood in a "Television Event No Parent Should Miss!"[3] His show, however, was not aired as the expected daytime segment, but was on primetime CBS. Titled *Family First* (incidentally also the title of his most recent book), this two-hour long show featured segments as provocative as families in crisis and the profile of a "serial killer" child, and as practical as potty training or raising a child's IQ score. The families featured on the show were representative of middle-class nuclear American families and Hollywood celebrities. To council these American families, Mc-Graw used sophisticated media technologies, including surveillance apparatus like the hidden camera, psychological and intelligence measurements, and personal narratives all flavored with his unique brand of home-spun commonsense.

This season is complemented by Montel Williams' debut as a prime-time host of *American Candidate*—Showtime's reality series featuring 10 men and women who are running their own personal campaign for president of the United States. Blending spectacle, political debate, parody, and civic responsibility, this program situates itself at the intersection of entertainment and information. *American Candidate* relies on the expertise of political "experts" who give advice on campaign management, polling, media coaching, and ad creation.

These examples taken together reflect how vulture culture functions—at once celebratory and corrective, political and innocuous—disseminating and reproducing particular kinds of knowledge, modes of living, and ways of being.

Vulture culture is best understood as the process by which the media scavenge the narratives, discourses, knowledges, and the everyday commonsense of our culture and present them back to us as information, spectacle, and entertainment. Vulture culture produces celebrity and relies on celebrity to promote itself. The fame of Winfrey, McGraw, and Williams is crystallized on these shows so much so that fame has become expertise. They have become the incredibly credible celebrities.

Vulture culture is a useful concept. It helps us understand the links between information, entertainment, and social, political, as well as economic institutions. These links exist on at least two levels. They are expressed within the content of the programs themselves through reliance on popular narratives, designated experts, and predetermined show formats. The implications of this sort of inquiry need to be made clear—the creation of topics, their repetition, as products and creators of discourse(s), have consequences for the ways we perceive and conceive our social worlds. Moreover, as we will see in the chapters of this book, the expert plays a defining role on these programs by at once legitimating the program topic, the hosts' perspective, and the networks' ideological leanings.

In vulture culture, these links are also established and reinforced through our consumption—as viewers and participants—of the values, norms, assumptions, and knowledges that our societal institutions produce and legitimize. For instance, in the *Family First* special, the discourse of parenting does not discuss what parenting is but rather promotes a particular kind of parenting: under the label of parenting, the program puts an emphasis on mothers' roles. Parenting becomes equated with mothering as it is the mothers in *Family First* who are charged with instituting the changes in their families. In addition, it should be noted that these are mothers within nuclear families. Furthermore, it masks the existing diversity within the family institution as America experiences it today. The program's overwhelming reliance on the traditional nuclear family with the exclusion of other family forms naturalizes and normalizes the heterosexual, two-parent household.

Vulture culture is also a useful concept because it helps us comprehend the mobilization of common sense around and through spectacle. Vulture culture is built on the spectacle. This is why talk shows are popular and central in this culture. Television talk shows are part of vulture culture because they promise the possibility of dialogue, communication and interaction between different members of society, yet they do so mostly through the laughs and tears, the ridicule and sarcasm of personal accounts of everyday lives. They prey on the anxieties, needs, and dreams of people. They feed on the triumphs of some and the tragedies of others; the pleasures and insecurities of many. Vulture culture seeks to awaken in many viewers the hidden voyeuristic impulses by continuously probing into the most intimate recesses of individuals' lives.

The culture of daytime television talk shows is part of vulture culture because, like many other media products, it seduces and betrays at the same time. Talk shows seduce us with ideas such as knowledge sharing, equality, social justice, and democratic values.

However, they betray us when they use these ideas to better position themselves in the competitive media market. Talk shows open the possibility of awareness and consciousness-raising but do so while entertaining through their tireless display of human frailty and follies.

SITUATING THE BOOK

This book approaches television talk shows as daily expressions of vulture culture and the logic that sustains it. We do not focus on daytime talk shows as decontextualized texts. We look at them as significant sites of knowledge production and consciousness construction within the context of modern-day America. We use the programs to initiate and encourage a critical debate on the complex processes involved in the production, transmission, and consumption of knowledge within this broader context of vulture culture.

The fact that talk shows focus on the confessional, capitalize on the "scandalous," and redraw the boundaries between the public and the private spheres of experience should not be cause enough to declare them "irrelevant" subjects of inquiry. There is a tendency in academic research and popular media to consider them "apolitical" spaces for the cultivation of deviance and mass circulation of irrational discourse.

The objective of this book is not to devise strategies for the recuperation of this popular media genre as a worthy subject of academic inquiry. Rather, the aim is to move the debate about and investigation of talk shows beyond the trappings of a bipolarity that either encourages an instinctive appreciation of the genre or urges its uncritical dismissal.

Until fairly recently the academic literature on talk shows was relatively sparse because researchers in the field of media and cultural studies have remained surprisingly reluctant to consider talk shows as an academically "legitimate" subject of inquiry. Existing literature falls within several broad categories:

- The history of talk shows and their position in the media culture exemplified in the work of Carbaugh, 1988; Munson, 1993, Shattuc, 1997.

- The assessment of audience preferences and perceptions to be found in the work of Priest, 1995, and Julie Engel Manga, 2003.

- The political implications of the public/private spaces and the role of the media in political discussion to be found in the work of Carpignano et al., 1990; McLaughlin, 1993; and, Livingstone and Lunt, 1994.

- Race, class, and gender on the talk shows in the work of Janice Peck, 1994; Masciarotte, 1991, Joshua Gamson, 1998.

- Issues of sensationalism and trash media in the work of Abt and Seesholtz, 1994; Abt and Mustazza, 1997

- Production of talk shows and behind the scene practices, as in the work of Laura Grindstaff, 2002.

Vulture Culture takes the talk show as a case study of the complex ways in which vulture culture operates, maintains, and reinvents itself. The book explores the intricate ways in which politics are woven in and through the production, transmission, and reception of this media genre. The concept of vulture culture allows us to situate *Oprah*'s car giveaway as something other than a generous gesture resulting from the harmonious marriage between *Oprah* and Pontiac.

One of the goals of this book, indeed, is to reveal how talk shows are important sites of knowledge production where the political is ever present at the level of content and form decisions as well as distribution and reception processes. We argue that if current discussions about talk shows have declared them "apolitical" spaces, it is mainly because the ambivalent nature of this media product has not been studied closely enough or stressed sufficiently enough.

Like most popular media products, talk shows contain both regressive and progressive impulses and, thus, operate between the shady zone of at least two contradictory and conflicting practices: the democratic urge to enfranchise and inform, on the one hand, and the commercial imperative to entertain and conform, on the other. Unlike other media products, however, talk shows celebrate the mixing of genres, thrive on the blurring of boundaries, and encourage contradictions as well as incongruity. Does this intrinsic ambivalence of talk shows, then, make them apolitical spaces? The arguments made in this book all point to the contrary. The dubious nature of talk shows might frustrate the need for an easy classification of the genre, but their elasticity does in no way preclude the many levels at which the political is manifested.

To advance the critical project of this book, we provide an analytical inquiry of talk shows in relation to questions of commodification, power struggle, and knowledge production, as well as the repositioning of patriarchy.

An investigation into the politics of daytime talk shows as a manifestation of vulture culture inevitably calls for a wide range of theories and approaches. In order to open a new space for reconceptualizing the pedagogy of talk shows, we draw on the overlapping concerns of political economy, cultural studies, and critical pedagogy.

I. THE POLITICAL ECONOMY OF TALK SHOWS

Vulture Culture provides an alternative project to current studies of talk shows which tend to focus on therapeutic discourse and audience reception. It is concerned not only with representative justice but also with issues of political economic concerns at all levels of analysis.

We utilize critical political economic analyses in this book to demonstrate how any single talk show hides multiple convergences and intersections of political and economic interests. These interests shape the content, format, topics, and the ideological parameters of the debates taking place on the shows. So it should come as no surprise, then, that the McGraw *Family First* defines parenthood with the help of other CBS celebrities such as Patricia Heaton and Amy Brenneman. What this tells us is that CBS uses the testimony of these actresses not only as real-life mothers but also as on-screen mothers of two of the most popular shows on the network (*Everybody Loves Raymond* and *Judging Amy*). The appeal to these famous figures helps McGraw construct the meaning of parenthood around the roles of the guilty working mothers while serving as free and further promotion of CBS programming.

Political economy is concerned with the study of "the social relations, particularly the power relations that mutually constitute the production, distribution, and consumption of resources."[4] Derived from the critique of capitalism, political economy focuses on media structures to explore the implications of ownership and control of the sources of information.

The question of the talk shows' ownership and control (chapter 2) is examined within the general political economy of communications. Discussion of concentrated ownership permits a better understanding of the roles played by corporate giants in shaping the production, regulation, and ideological orientation of talk shows. This focus leads to the investigation of the processes of commodification and commercialization in which talk shows are embedded (chapter 3). Commercialization is discussed at three main levels— content, audience, and labor—to reveal the insidious routes talk shows take to establish the vulture culture.

The political economic approach is a necessary step towards comprehending the effects that the capitalist system has on its cultural/media products and the everyday lives of the consumers. *Vulture Culture* seeks to further reveal the intersections between the political economy of talk shows with issues of race, class, gender, sexual orientation, as well as democratic spaces.

2. CRITICAL CULTURAL *STUDIES* AND CRITICAL PEDAGOGY

A critical cultural approach explores the different ways in which media texts support and/or undermine the dominant ideologies and value systems in society. This approach enables us to rupture the boundaries of vulture culture by addressing questions of power, representation, ideology, and resistance. These questions are shaped and defined by various institutions within vulture culture and are woven in the production, content, and consumption of daytime television talk shows. To understand and uncover the meanings and implications of the talk shows, we take a multiperspectival[5] and multicultural[6] position in the critical cultural studies tradition. This approach involves using a wide range of insights such as Marxism, feminism, and discourse analysis.

This multiperspectival approach to the shows will reveal how the daily programs contain different types of power struggles at the individual and collective level, as well as struggles over values, behaviors, and choice of lifestyles. We will also show how popular media products such as the shows actually work against any potential democratic impulses the programs might be suggesting. For instance, when Jerry Springer started his show, he was trying to meet the needs of a specific population—the young adult. Opening a public space to often marginalized and excluded voices could be seen as a democratic gesture of inclusion. However, Jerry Springer prioritized the style over substance, the spectacle over meaningful articulation of real-life concerns. The spectacle certainly has made the programs profitable, kept the networks happy, and the viewers entertained, but in the process it has erased the possibility of democratic debate and exchange. In this book, a critical cultural approach is used in connection with critical political economy and critical pedagogy to flesh out the economic, political, and cultural implications of the vulture culture that sustains the talk shows.

The critical cultural approach of the book is informed by the concerns of critical pedagogy. Critical pedagogy in general attends to questions of knowledge production and power dynamics involved in constructing, transmitting, and imparting information. In investigating these processes, critical pedagogy provides a better understanding of the mechanisms that shape and influence consciousness. Further, critical pedagogy places special focus on questions of justice, equality, and democracy with respects to different sites of learning and knowledge building which are not limited to traditional school systems.[7] Whether it is in the line at the grocery store or in the corridors of a high school, talk shows have become a part of everyday conversation providing everyday information exchange. A critical pedagogical approach recognizes that viewers of media products in general and talk shows in particular use these programs as sites of learning. For example, the viewers of *Dr. Phil* have grown to expect to find useful or practical tips such as losing weight, solving family tensions, and money management. And yet, critical pedagogy does not stop at just recognizing the use made of the information. It points to a higher level at which the programs' information shapes our perspectives and understanding of social problems and cultivates the norms and values by which we comprehend our social world.

Chapter 4 looks at the types of knowledges produced, promoted, and privileged by the shows. "Expert" and "commonsense" knowledge is examined in relation to issues of power, identity, and stardom as they are played in these daily programs. Indeed, so much in the talk shows is precisely about "playing," staging, and performing (chapter 5). Talk shows' reliance on the spectacle formula has turned them into what we call "the electronic carnival": a truncated reproduction of the earlier popular form of festivity.

Chapters 6 and 7 provide an in-depth reading of ways in which gender issues intersect with power, knowledge, and discipline on the daily programs. They look at the various institutions and disciplinary knowledges that shape, define, confine, and control women's bodies and behavior. A close analysis of the teen-out-of-control episodes demonstrates how discourses on women and girls are intricately intertwined with the criminal justice and welfare systems. The recent proliferation of paternity shows, the focus of chapter 7, confirms that the paternity test as a scientific and legal document serves the interest of capitalism while prescribing normative roles for motherhood and fatherhood.

Taken together, these chapters address the overlapping labels of race, class, gender, and sexuality, providing insight as to how each of these elements is inherently connected to a larger conservative agenda-setting politics.

Vulture Culture argues that television talk shows constitute an example of "non-traditional" sites of learning and knowledge production, and as such the programs cannot be simply declared irrelevant. Our task, therefore, is to critically analyze the various types of knowledge created in these media sites, the diverse strategies that permit such a creation as well as the multiple effects they might or might not intend to have. Because of the considerable popularity and visibility of these programs, the book encourages educators, cultural workers, and concerned citizens to develop a more critical awareness towards the talk shows and the larger environment within which they operate. This environment is precisely what we have termed vulture culture.

2

The Commercial Logic of Vulture Culture:

How Corporate Media Shape Talk Show Culture

D aytime television talk shows are important cultural spaces in which social, cultural, economic, and political issues are debated, identity narratives are told, and insights into the fears and desires of society are created and reproduced. As mentioned in the introductory chapter, these shows are not "innocent" cultural expressions. Rather, they are discursive constructs and commercial products of a culture industry whose goal it is to create and sell cultural expression for profit. They are part of the corporate media system and its particular axis of vulture culture.

Rather than leave this statement at the mere assumptive level, this chapter interrogates the relationship between talk shows and larger institutional commercial television systems, corporate media structures, and general political and economic institutions. Thus, analyzing talk shows from a political economic perspective allows us to understand some of the main characteristics of vulture culture, the driving force behind this culture and the logic that sustains it, for talk shows are among the most popular expressions and manifestations of vulture culture. This chapter demonstrates that if a serious understanding of cultural production is to take place, then the economics of television must be seriously investigated. This is so because economic imperatives impact not only *whose* cultural products will be produced, but also *which* ones will be produced, in whose interest and to whose benefit and detriment.[1] The chapter also addresses the impact that the corporate/commercial television industry has on the production, distribution, and content of daytime talk shows.

TALK SHOWS IN CONTEXT

In order to understand talk shows as a media product, we need to understand the circumstances of their production. We argue that talk shows, as expressions of vulture culture, are constantly reinventing themselves. This reinvention reflects the ways capitalist economic institutions change to suit the emerging social, political, and economic contexts that such institutions inherently shape. However, situating the talk show in a particular historical moment proves to be more difficult than one might imagine, as the characterization of today's economic context is under dispute. Is this a world of post-Fordism, or late capitalism? Have we just entered a new phase of capitalism, or does the current climate simply reflect the continuously evolving symptoms of capitalism?

Douglas Kellner calls the current state of capitalism a new "configuration of constellations." Rather than a new stage, he sees these constellations as forms of both progress and domination.[2] He argues that the classical capitalist processes continue; the roles of automation, computers, and technology now "parallel the role of human labor power, mechanization and machines in earlier eras of capitalism, while producing as well new modes of societal organization and forms of culture and everyday life."[3] Techno-capitalism "is thus characterized by a new synthesis of technology and capitalist social relations and by production of new techno-commodities and techno-culture."[4] This conception of late capitalism helps us understand the persistency with which capitalism shifts to enclose new technologies. In a techno-capitalist context, communications systems play a vital role in the spread of capitalism as well as the production of techno-commodities and techno-culture. A survey of talk shows suggests that indeed they provide an ideal site for the examination of techno-capitalism, especially around episodes which feature paternity. These age old stories are reinvigorated by the use of DNA tests (a techno-commodity) to create a new version of this spectacular drama (a new techno-culture). In this way, revision, reinvention, and innovation are hallmarks of both vulture culture and capitalism.

To further understand the talk show in context, we must understand it as a commodity. America is a society in which cultural expression—including talk shows—often exist as a commodity to be owned, bought, and sold according to the logic of capital, by private interests, for profit. The fact of talk show ownership may seem obvious: *of course* a talk show is owned by someone (or something). But this is not a natural fact, one that exists out of pure and natural causes. Rather, ownership is a historically determined concept. Who owns what is determined by society's values and decisions about the allocation of social, cultural, political, economic, environmental, and other resources. When left unanalyzed, the way that cultural resources are allocated seemingly becomes normalized, as though it were a natural and socially acceptable way of distributing resources. But as described, ownership is anything but a natural process. Instead, we need to think of ownership as an exercise and expression of power that has ideological, economic, and political implications. The talk show as a cultural commodity is bound up with questions of cultural ownership, to be addressed more fully below.

Understanding talk shows as both situated in a techno-capitalist context and as positioned historically through processes of ownership necessitates a revisiting of some of the key ideas and concepts of political economy. This will allow us not only to see how the talk show is situated as a commodity in the "free market" system, but it will give us the language and the concepts we need to foreground the political and economic forces that work to constrict the talk shows' potential as a democratic form of communication and consign it to the realm of vulture culture.

REVISITING CAPITAL

It is important to review some elements of capitalism, economics, and political economy in order to better understand the systems through which talk shows function. According to classical economics, capitalism functions when producers of commodities compete to sell their products to consumers in a "free market" of exchange; supply and demand operate in equilibrium. This view suggests that the "free market" is regulated by a set of market checks and balances, ideally with little or no government intervention. The underlying assumption is that each competitor has equal access to truthful and complete information, as well as equal access to markets and customers. The goal in the capitalist system is the accumulation of capital and wealth. The strategies used to accomplish accumulation include working in economies of scale and scope, branding, integration, diversification, globalization or market expansion, joint ventures, or partnerships.

However, critics of classical economics have documented how a combination of political and economic interests works to prevent competition and encourage monopoly. The critical political economy approach takes to task these assumptions of classical economics and neo-liberal economics, arguing that power is at play in oppressive ways. Because some people own the means and rights to reproduce social life, and others must sell their labor power, political economists argue that capitalism is not a value-free system but one that is based on unequal power relations. Further, political economists analyze the connections between economic, political, and communicative process, and do not assume a de-politicized economic sphere. Finally, political economy asserts that moral philosophy, praxis, social totality, and historical context are important elements to an analysis of political economy of communications.[5] These political economic concerns are vital to understanding the scavenging tendencies of vulture culture, in that the injustices perpetuated in the name of talk shows' free-market economics are uncovered and critiqued.

SPATIALIZATION

A crucial theoretical concept for a critical political economy of talk shows is "spatialization." Simply put, spatialization is a way of considering the dynamic and ever changing issues of media ownership and control in a larger capitalist context. Mosco defines this concept as "the process of overcoming the constraints of space and time in social life."[6] In other words, the concept of spatialization allows us to understand how capital finds faster and more efficient ways to grow over larger areas in less time. As capitalistic processes spread over and transform space and time, two specific points of continued analysis should include: First, the expansion of corporate power into communications institutions; and second, the forms and ways of corporate concentration.[7]

In the current techno-capitalist political economic context, we can see the role of spatialization in investigating an individual talk show. One can determine who owns the show, as well as who has control over decisions regarding the show—in terms of its production and content, its price on the market, its profit margins, etc. The expansion of corporate power over communication and cultural institutions can be seen in questions of copyright. Ownership of television programs is determined by copyright.[8] Copyright, a form of intellectual property, grants the owner control over the reproduction, display/exhibition, sale, and profit for that literary expression and thus limits the cultural usage of the item.[9] The owners of talk shows can be analyzed in light of their participation in concentrated media markets.

Furthermore, spatialization points to the importance of understanding media mergers in terms of corporate concentration. Right now, in the merger-friendly neo-liberal political and economic climate, the major Hollywood studios and television networks are becoming more integrated. Each network and major studio are part of the handful of companies that own most other television programs, channels, and other media and cultural products; the main players in television are the same players in every other media industry or genre, their power depending on the strength of their holdings in any particular market.

By analyzing the companies that own talk shows, one can readily see how talk show owners are intertwined with one another and other media products. As a brief example, *Live With Regis & Kelly* is owned by Disney, one of the major Hollywood studios that produces and distributes children's films (from *Dumbo* to *Aladdin* to *Princess Diaries*), as well as general films through subsidiaries Touchstone Pictures and Miramax Films. Disney also owns cable's Disney Channel, the Playhouse Disney kids' television programs, the ABC Network, Disney Stores, Disneyland, Disney World, Disney Cruise Lines, and more.[10] Thus, *Live* is not merely an individual talk show, but part of a web of Disney holdings.

Even though intuitively, one might assume that the television networks (ABC, NBC, CBS, and the newer Fox, WB, UPN) have more power in television, the major Hollywood studios have been granted the greatest access to television production and distribution, due to specific regulatory measures meant to bolster competition in television (especially through the Financial Interest and Syndication Rules, or Fin-Syn). On the other hand, the primary television networks have controlled exhibition through broadcasting, including network-owned and -operated stations (O&O) and a massive system of network affiliates spattered across the country, most noticeably through the power to pick up or contract to exhibit a syndicated talk show on their stations. These processes have been enabled by state and market regulation. In particular, state "de"regulation of the marketplace (still a form of regulation) ultimately enables the spread of companies into different sectors of the communications industry. This is clearly evidenced by the merger craze in the mid-1980s (continuing to today), as President Ronald Reagan's economic plans deregulated the media industry in a way that encouraged such mergers. This sentiment is echoed in the early years of the twenty-first century, in the Federal Communications Commission (FCC) headed by Michael Powell, son of the former Secretary of State Colin Powell.

The spatialization of talk shows is thus an important element of vulture culture. The specifics of government and market regulation facilitating corporate cultural control of media culture, including talk shows, establishes a web of corporate culture that dominates the media landscape.

INTEGRATION

Within this political economic climate, talk shows' individual markets should be analyzed. The television talk show industry occupies three markets in the media landscape: the talk show production, its distribution, and its exhibition or airing on television. Tactics of techno-capital accumulation are evident in each of these markets. One of the central accumulation tactics is integration, or the control of multiple markets by one corporation. This allows media corporations to reduce costs and maximize profits by creating monopolistic control. Two types of integration can be described: vertical integration exists when a company owns the companies or products that are used to make, distribute, or utilize their product, while horizontal integration involves owning multiple companies or products operating in the same market. Both horizontal and vertical integration will be discussed by market.

Production consists of planning the content (including hiring the host, coming up with topics, screening guests, bringing in and warming up audiences), constructing and decorating the studio set, taping/filming, and editing the program. Essentially, production companies must finance the entire show, from paying talent and production crews, to buying cameras, microphones, lights, tape, and wardrobes, flying guests to the show and paying for their food and hotel, renting or owning studio space and postproduction editing facilities. The talk show production market includes a number of different companies creating the end-product show that we watch on TV. Some of these producers are independent and not owned by a large corporate conglomerate parent, such as KidRo, Rosie O'Donnell's production company and the producer of her show, MoPo—Maury Povich's own company, or Harpo Entertainment, Oprah Winfrey's production company, producing *Oprah* and co-producing, with Viacom's Paramount, *Dr. Phil*. Other production companies are either traditional television networks such as ABC Television, producers of *The View*, or Hollywood film studios such as Warner Bros. (owned by Time Warner), which produces *The Jenny Jones Show*; Telepictures, also owned by Time Warner, co-produces *The Larry Elder Show* and *The Ellen DeGeneres Show*.

The second television market is distribution—the deal that the production company makes with another company, which will reproduce or copy the show *en masse* and relay the program, via videocassette or satellite broadcast, to stations and thus be aired and seen by viewers. Distribution, while seemingly mundane and a matter of simple logistics, has increasingly become the key market in electronic entertainment. In fact, it is taken into account at the level of production, as a means to finance production.[11] Distribution companies, thus, are most often part of the large media conglomerates, the companies that have market power, brand recognition, access to exhibition markets, and money available. Talk show power is concentrated at the distributive level as well—for example, StudiosUSA Television Distribution (a subsidiary of Universal/Vivendi) distributes *Maury*, *Jerry Springer*, and *Sally Jesse Raphael*, before the company canceled her show.

The third and final television market is exhibition. As a point of comparison, in film, an exhibition venue would be the movie theater that shows a film. In television, the exhibitor of a show is either an entire broadcast or cable network (NBC, ABC, CBS, Fox, WB, UPN—the broadcast networks; or Disney Channel, Comedy Central, HBO, Nickelodeon, etc.—cable channels), or an individual television station that contracts with cable companies and broadcast networks to show programming. In the case of talk shows and some other daytime programming, individual stations play a large role. Syndicated programming is bought independently of networks and their affiliated stations, and is aired on whichever local stations buy them. Talk shows, with a few exceptions, operate this way, in syndication. For example, *The Ellen DeGeneres Show* is sold in syndication to interested local stations across the country; however, it is also aired on all NBC owned-and-operated stations, and on Oxygen network on cable.[12] Syndication is in contrast with network affiliation, which requires affiliate stations to air the entire slate of network programming. Typically the prime-time shows are only broadcast by affiliates (e.g., *Law & Order* on NBC affiliate stations) but a syndicated program like *Oprah* can be bought by an NBC or Fox affiliate, or an independent station.

To make talk show integration concrete, a few additional examples are in order. Disney owns studio production facilities, distribution firms, cable channels, and a television network.[13] Thus, they produce their own talk show (*The View*), distribute it (through their distribution firm Buena Vista) and air it on ABC, their own network. They are therefore a vertically integrated company. By owning properties at each point in the chain of production, distribution, and exhibition, they completely control a product and its profits. There is no "free market" in this instance—no other companies are competing to contract for *The View's* distribution, as Disney would rather keep the profits in-house than to allow another network to get a piece of the pie. Because of integration, smaller companies are not hired, as the major media companies monopolize the market. Anti-competitive practices occur also with horizontal integration. For example, the talk show distribution market would be more competitive if every distribution company had a chance to bid for programs, and no one or small handful of companies dominated the market. However, we see a small group of corporations dominating talk show production. Paramount (owned by Viacom) distributes Montel Williams, and Viacom's horizontally integrated King World distributes *Oprah* and *Dr. Phil*. Warner Bros. distributes *The Ellen DeGeneres Show* and *The Larry Elder Show* (in addition to co-producing *Elder*), making Time Warner both horizontally and vertically integrated. A final example: as discussed, Vivendi distributes *Jerry Springer* and *Maury*. As we can see, this is not a competitive market, but rather an oligopoly, as a few companies control a majority of the market, thus providing an unfair advantage to these large, powerful companies. Vertical and horizontal integration prohibit competition, enable consolidation of power and control, and lead to concentrated markets, which will be discussed momentarily. As an element of vulture culture, integration has a direct impact on talk shows in that a handful of companies profit from these programs, scavenging the cultural potential of talk for the material gain of a select few.

DIVERSIFICATION

Diversification is another accumulation strategy with monopolistic implications. Diversification allows corporations to spread out their economic risks by owning a variety of different companies and products, some of which may have slow growth and profit periods while others prosper greatly. This practice attempts to prevent the whole corporation from failing if one of its industries has difficulties; the parent company does not rely simply on one sector for their survival. A parent company that primarily focuses on film distribution but also owns music labels, television channels, video chains, televisual and computer hardware and retail outlets is a diversified corporation. Such a conglomerate can be formed by mergers and acquisitions, as are rampant in the media industry. Antitrust laws created to prevent such behemoth companies are largely ignored, as seen in the AOL Time Warner merger, or the Viacom/CBS merger.[14] Each of the major media companies is a diversified company. A good example is Time Warner, which owns television and film studios, music recording labels, the WB television network, cable system operations, cable channels—including CNN, all of AOL and its holdings, and numerous magazines, including *Time*.[15] The company is notorious for buying other companies, yet ironically was bought by AOL to create a giant media octopus. Time Warner is involved in the talk show industry through its co-production of *The Larry Elder Show*, and its distribution of *Elder* and *Ellen DeGeneres*. These talk shows, as part of the larger media conglomerate of Time Warner, are protected by the vast holdings of the parent company, in that their economic success or potential failure is buffered by the rest of Time Warner's profit-making sectors and products.

Companies can be considered either "core" or "periphery" firms. Core firms are the Fortune 500 companies, and all other companies are considered peripheral. Most of the major media companies that produce, distribute and exhibit talk shows are part of a core firm, which allows them access to markets and capital and a lower degree of risk as being backed by such a large core company. Over the years 1985 to 1995, "the share that Fortune 500 leading corporations represent[ed] of gross national product rose from 44 percent to 63 percent."[16] The implication of this shift is that larger corporations are gaining wealth and power, whereas other interests are losing power. Because of capitalism's tendency toward concentration, then, cultural life increasingly becomes monopolized and corporations have more control over cultural spaces, artifacts, and expression. Ben Bagdikian's *Media Monopoly* (2000) details the decreasing number of media companies over time, and more importantly the amount of market control being shared by fewer and fewer companies. The strategies of diversification and integration have led to this concentrated marketplace, which has produced a strong television oligopoly, consisting of Time Warner (owner of Warner Bros., WB), News Corp. (owner of Fox), Viacom (owner of Paramount, UPN, CBS), and Universal (owned by Vivendi, a French utility company).

Diversification, justified by economic stability, is a strategy that results in media conglomerates operating in numerous markets. As such, our telephone operator becomes our cable operator becomes our electric utility provider.[17] These cross-industrial moves are enabled by market and policy, not natural processes, and result in a climate in which a few companies control an inordinate amount of our cultural life. Talk shows are no exception. For example, Dr. Phil's program is co-produced by Paramount, a company owned by Viacom, and distributed by King World, another Viacom subsidiary. His primetime special, *Family First*, aired on CBS, is also owned by

Viacom. The book, *Family First*, is published by Simon & Schuster, again owned by Viacom. In addition, special guests on the show included Amy Brenneman and Patricia Heaton, stars of CBS prime-time programming (*Judging Amy* and *Everybody Loves Raymond*). In this way, Viacom flexes its horizontal, vertical, and diversified muscles in order to showcase McGraw as a branded commodity. This action prevents smaller names or new programming from receiving attention. While the company may argue that McGraw is popular and he wouldn't be successful if consumers didn't use his products, it can be countered that this system is creating a lack of choice, not responding to consumer desire and demand. While it is no doubt that McGraw touches on viewers' fears and desires, it is also important to remember that the show, special, and book, are components of a marketing scheme for Viacom to solidify market power.

INDUSTRIALIZATION

Talk shows operate within the landscape of the mass media as media products and cultural commodities; as such they are industrialized.[18] Industrialization involves the mass production and capitalist control over cultural commodity production and distribution. Thus, culture is *commodified* and follows capitalism's industrialized processes of expansion, diversification, integration, concentration, and tendencies toward monopoly control over markets and culture.[19] Given the industrial nature of television, and its tendencies toward concentration, it should be no wonder that there are a handful of companies involved in the industry.

In this context, the spatialized web of talk show ownership has been created by major corporate conglomerate strategies of integration and diversification. In such a situation, vulture culture is founded on a constricted notion of culture, one in which a handful of organizations create, disseminate, and profit from television talk. As part of international conglomerates, talk shows function as cheap filler that pale in both production costs and profit potential when compared to billion-dollar blockbusters. However, talk shows' place in media culture is one of forbidden delight. The contradictions between the seeming unimportance of talk shows and their important business, cultural, and political implications are not lost on our analysis of vulture culture.

CONCENTRATION OF POWER: A SOCIAL PROBLEM

The extension of corporate power into the communications industry is thus not totally separate from the issue of concentration of wealth and power. Concentrated power is a social problem, not simply a media or business issue. In 2001, the richest 10% of the U.S. population held 60% of the wealth, while the lower 90% of the population held only 30%.[20] The CEOs of the Fortune 500 media companies are part of the wealthy elite, while those who watch talk shows are generally part of the 90%. Thus, we already see a difference in lifestyles, goals, and problems between the audience and the talk show producers. In the upper echelons of society, corporate officers, financiers, and government officials are linked together through their class status as elites. More specifically, elites are integrated through direct and indirect interlocking boards of directors as well as networking or shared experiences or administrative positions at ivy league schools, seats on think tanks and in government lobbies. Direct interlocks, when a corporation's board of directors seats a board member from another large corporation, and indirect interlocks—those in which a third company's board contains members of two other corporations' boards—help the elite class scratch each other's backs in an old-boys club mentality. They are able to give their elite friends or associates advantages when profitable situations arise.[21]

Such an incestuous corporate elite merely scratches the surface of concentration, however. Concentration through ownership is illustrated in a number of ways. Traditionally, analyses of concentration have examined the market share a company holds, which is an issue of antitrust. Two other important projects for examining concentration are to look at the size of a company in comparison to other companies, and to analyze product diversity in the market.[22] Concentrated ownership creates large amounts of power through corporations' exploitation of their market share, company size and ability to create either diverse or homogenous products.

The disparity in social wealth and power is important to a discussion of vulture culture insofar as vulture culture feeds on, contributes to, and masks these disparities. Talk shows as vulture culture are integral components of media conglomerates that are some of the largest companies in the world, impacting global cultures and communities.

Further, the content of talk shows often sidesteps issues of disparity while simultaneously publicly displaying that same disparity.

SERVING THE PUBLIC GOOD

The implications consolidated corporate talk show ownership have for vulture culture begins with the discussion of a contextual policy issue. By legal mandate, through The Communications Act of 1934, the broadcasting system is endowed with the dictum to serve a larger public interest, convenience, and necessity. As such, television is supposedly held accountable by and "owned" by the public. As an integral component of the so-called political democracy, communications systems are therefore theoretically democratic tools, used to inform citizens in order for them to be valuable participants in democratic governments or other institutions. Media should, according to policy, offer a space where debates, serious news stories, labor issues, economic concerns, and cultural events are displayed and discussed, with the goal of supplying information and entertainment in democratic and meaningful ways.[23] Coming from a political economy perspective, a moral philosophy for social justice and equality is held up as the standard for operation. As we have seen, this ideal is not currently being realized. The logic of capital accumulation and the logic of legitimation work to erase political economic ties and structures that shape talk show and other media content.[24] In so doing, capitalist media legitimate and reify ideologies and structures that support the reproduction of a capitalist society, commercial television and a consumer culture. This hegemony is oppressive and inhibits democratic functions supposedly endowed to the airwaves as an integral part of the public sphere.[25]

In constraining television's potential democratic impulses yet claiming that democracy does in fact exist vis-à-vis the so-called marketplace of ideas, talk shows, and media in general, contribute to a marginalization and displacement of alternative voices.[26] Again, the marginlization is linked to market structure. The concentrated, monopolistic market leads to politically and economically imbalanced media, for as ownership is concentrated, small independent media are reduced, and real choice and debate are diminished. Cultural and social diversity and multiculturalism are also lost as the corporate-owned media protect their economic interests, avoiding what they perceive as risks in structure and content. In a commercial system, media and culture are seen as commodities; exchange value is key in attracting both advertisers and audiences, and content is created so as not to offend either. In this vein, *The Ellen DeGeneres Show* is rife with celebrity guests and light musical and comic fare, rather than a serious discussion of gay and lesbian politics. Media products therefore become standardized and synergized, as seen in the film and television industry with sequels, spin-offs, star system, soundtracks and toys—television talk shows are made to sell (see chapter 3).

Further, the trend toward global concentration calls for world marketability of big-budget films which need to sell worldwide to recover their fixed costs, coupled with low-production-cost programming, such as talk shows, that can act as filler. Talk shows' popularity domestically and abroad can be contributed not only to their appeal to the audience, but because of the supply-side issue of their low production costs. Because they have a volunteer studio audience and volunteer guests, the celebrity host and the production staff are the only labor costs. Much of the camera and editing work is done by contract workers who ride the "talk show circuit" weekly, putting in time wherever work is available. These practices undermine creativity and diversity of domestic products, which tend to focus on cheap, easily translatable talk show spectacle rather than creative, detailed, and nuanced scripts or debates.

COMMERCIALIZATION, GLOBALIZATION, AND THE PUBLIC SPHERE

With ever-increasing speeds of development and capture of new technologies, which are robbed of any potential democratic impulses by corporate monopolization, the combined forces of commercialization and globalization work to threaten the possibility of a consequential public sphere. As more spheres, whether physical, mediated or abstract fall under control of corporate commercial imperatives, the marketplace shapes the content of those spheres, and thus severely constrains democratic possibilities. Talk shows, with visions of populist participation, which purport to level power relations and provide arenas for multiple and different voices, merely concern themselves with making a profit. As will be discussed in chapter 3, commodification shapes the possibilities for talk show public spheres, so much so that within the confines of a commercial system, a viable public is arguably impossible.

Lisa McLaughlin's article concerning talk shows and public spheres elucidates the problems of corporate colonization of common sense (to be discussed fully in chapter 5), concluding that "the problem with locating talk television at the crossroad of collective practices that constitute public sphere activity is that it is no more empowering or revolutionary than locating Madonna on the cutting edge of feminism."[27] In other words, celebratory authors writing about talk shows as valid public spheres, make the same analytical omissions as scholars who celebrate Madonna's feminist "resistance" through identity appropriation in ignoring issues of appropriation/ accommodation, commodification, commercialization, and lack of real resistance to oppressive patriarchal structures.

Additionally, globalization of transnational media conglomerates enables further capitalist accumulation as talk shows are exported to other countries: *Ricki Lake* is available in Scotland, and *Oprah* is exported to 106 international markets. The talk show genre has influenced local development of talk shows in other countries. For example, a Russian talk show called *About It* is directly influenced by American talk shows, focusing on confrontational topics. The host, Yelena Khanga, who is half African and part American and Russian, wears a blonde wig on the show; her producers tried to get her to wear blue contact lenses, but she refused. On the show, she acts as a therapist and psychologist, although her educational training is in journalism, like many American talk show hosts.[28] Considering the "break-up of Communist Eastern Europe and the accelerating restoration of capitalist forms and practices there" and elsewhere in the former Soviet Union, the trend to globalization[29] illustrates the growth of transnational corporate power, which limits alternative media systems and prevents democratic, participatory structures from forming.

Thus, issues of cultural imperialism must be considered in the attempt to understand the changing global and transnational media systems and their impact on local and global cultures. Talk shows' role in cultural imperialism must be introduced in order to understand how the genre plays an integral role in the economic, cultural, and political exploitation of many international cultures simply because of its cheap production costs. Hence, vulture culture is exported globally. The general effects of such vulture culture imperialism are the cultivation of the culture of consumerism, a homogenization of local media products, the proliferation of commodified and non-participatory public spheres, and a prevention of real democracy.[30]

Each of these expansionist processes illustrates the tendency for capital to concentrate into larger and more controlling companies, with the constraining effects including the invisibility of independent producers, a stifling of creativity, the lack of alternatives, and restricted access to the public airwaves. In order to compete with the "big boys,"[31] creativity must be co-opted in order for a show to become more marketable and advertising friendly—to support an inviting business atmosphere. Formulas become more important as shows which have proven their ability to garnish ratings are copied, illustrated by the long chain of early shows copying *Donahue*, and the late 1990s string of *Rosie*-alikes: *The Howie Mandel Show*, *The View*, *The Other Half*, and into the twenty-first century with *Wayne Brady* and *The Ellen DeGeneres Show*. These post-*Rosie* shows use similar sets, topics, language and rhetoric, graphics, and ultimately cater to an advertising-friendly format which uses celebrity interviews, baking, and other innocuous fanfare. As long as the political economic forces remain in control, we can expect these hegemonic trends to continue.

The questions raised in this chapter circle around the concept of power. Although the paradigm of media power has changed over the years, in this book, we are not arguing that media power lies simply in the institutions, nor solely with the viewers. Rather, we posit that a combination of forces work together to create the power that talk shows have in our society.[32] With this in mind, though, we might add that not all forces are created equal; the power of a multi-million dollar conglomerate is in most instances stronger than that of a single talk show viewer. This chapter has delved into questions of media power through the central concept of spatialization. Without the logic of capital and accumulation, the intersection of forces that create talk shows would not be strong enough to support such an exploitative and oppressive genre. The ways in which corporations use their political economic power to control the talk show genre and the rest of the media landscape have been uncovered and discussed with the goal of identifying specific instances of uneven power in the talk show industry. The reasons why concentrated ownership exists, as well as the role corporate giants play in shaping the production and regulation, both market and government, of talk shows have been integral to the discussion. We also must look towards how these institutional structures relate to the content of the shows themselves and how these structures work with and against viewers' interpretations in a hermeneutical relationship. These issues are taken up in the next chapter.

3

THE COMMODIFICATION
OF TALK SHOW CULTURE

W hen we watch talk shows, we rarely think of the programs
in terms of commodification. When we do, we are quite
likely to limit our thinking about commodification to the
somewhat obvious commercial aspects of the program, such as the
commercial breaks and the "this program has been sponsored by"
quips at the end of the program. This chapter will push questions
about commodification further, into a more in-depth analysis of this
process.

UNDERSTANDING COMMODIFICATION

This chapter will discuss three commodification processes and their
effects on talk shows as a media product. The commodities themselves
include content, audience, and labor; each will be addressed. Further,
commodification is clearly linked to commercialism and advertising.
Twentieth-century capitalism evolved the use of advertising to
legitimate its production practices by creating lacks that products
could fill, and creating brands to which consumers may become
loyal. Thus, advertising helps encourage consumerism, with the
commodity at its heart. The goal of this chapter is to examine
how talk shows play a role in these legitimation and accumulation
processes through commodification.

Marx's discussion of commodities can help us understand the
cult of the commodity form in capitalism. He has written that in a

capitalist structure, the commodity has taken on a life of its own. As he clearly puts it:

> [T]he relations connecting the labour of one individual with that of the rest appear, not as direct social relations between individuals at work, but as what they really are, material relations between persons and social relations between things. It is only by being exchanged that the products of labour acquire, as values, one uniform social status, distinct from their varied forms of existence as objects of utility.[1]

Commodity fetishism allows social relations to be concealed, as the fetish "attaches itself to the products of labour, so soon as they are produced as commodities, and which is therefore inseparable from the production of commodities."[2] Thus, the commodification process defines the process of transforming use values—the practical value of something in one's life—into exchange values, the dollar value of a product. By "transforming products whose value is determined by their ability to meet individual and social needs into products whose value is set by what they can bring in the marketplace,"[3] commodification removes products from a more meaningful social context into one that primarily benefits businesses and the ideology of "free market" values. A culture based on commodity fetishism, then, values the exchange value of a cultural product as much if not more than the pleasures that it produces. Since processes of commodification are perhaps the central forces that propel capitalist expansion into more and varied realms of society, we can argue, then, that commodification should be central to a political economic analysis of talk shows.[4]

CONTENT AS COMMODITY

Looking at talk show content through a political economic lens, we can closely analyze the workings and material effects of the central capitalist process of commodification.[5] Talk shows' content may be studied as commodity at three interrelated sites:

1. The produced program as spectacle

2. The commodified "problems" of the produced show

3. The intertextual commodities used as consumer remedies to commodified aspects of the show

With respect to the first level of commodification, the talk show as a market product constitutes part of the "culture industry." Television talk show programs, in the broadest terms, exist as marketable commodities, usually syndicated, that are bought and sold in the communications marketplace, a process outlined in the previous chapter. These talk shows function as spectacles in that their visual imagery constitutes vulture spectacle, meant to entertain, amuse, and invite participation from the largest segment of audiences and viewers. By garnering a large market share, a show can command more advertising dollars—as share guides ad prices: an advertising dollar spent to reach 100,000 viewers is ten times more efficient than one spent to reach 10,000 viewers.

The filmed spectacle/program comprises three types of commodities: confrontation, information, and celebrities. Each of these is a marketable good insofar as each theme competes for popularity and further exposure on the spectacle/show. These second-order commodities can be conceived of as problems that need solutions— the solution being the third-order intertextual commodities. Information problems are solved with useful products; confrontation issues with therapists' books and make-over tie-ins; celebrities with films, TV shows, books, and music. No matter the problem, buy your way to a solution today (available at K-Mart for $9.99, while supplies last!).

Intertexts are created as products and texts combine in such a way that "text, intertext, and audiences are simultaneously commodity, product line, and consumer."[6] This process, popularly employed in films, enables talk show commercialization through advertising, product placement, tie-ins, and merchandising.[7] In contrast to films, which have event-like releases hyped with consumer products, television talk shows are serial, featuring different guests daily. Talk shows do not have momentous build-up and concentrated marketing schemata as film tie-ins and intertextuality, such as that found with *Batman*[8] or Disney films.[9] The nature of their tie-ins, though, is longer-term and focused on the celebrity host and guests. Much like the product placement of AOL in *You've Got Mail* or Wilson volleyballs in *Cast Away*, talk shows often function as hour-long infomercials for films, books, albums, videos, fashion, cosmetics, diapers, or politicians. Unlike movies, talk shows do not have high production costs and production value used to reel in viewers and advertisers with promises of special effects. So, producers construct an overarching theme that includes tie-ins and product-showcasing in order to imbibe in a culture of consumerism, appeal to a consumer-oriented audience, and obtain free products and more advertising money. An example is *Oprah's* "Favorite Things" episodes, in which she presents a spectacle of products that she loves. The episode focuses on the legitimation of these products, as well as their dispersal to the studio audience. With some film and book concepts, the same companies that produce or distribute blockbuster films produce or distribute talk shows. Through diversification, these production and distribution companies spread out their risks into different types of media—they can enjoy fruitful seasons in any one medium while being able to recoup losses or justify lesser earnings in their other products. Perhaps more importantly, they can exploit each medium through monopoly power to advertise for the others. Thus, by contrasting the tactics of ubiquitous consumerism and event consumerism, one can understand the rationalization behind talk shows' processes of commodification and pedagogies of consumerism.

Many talk show episodes center around the commodification of useful information on becoming a better consumer. Such topics as the year's best Christmas gifts, how to apply "natural-looking" make-up, how to get the best airline deals, what one needs to know about the cleanliness of public shopping mall toilets, and which boutiques to shop at for the season's newest bathing suit fashions (not to mention how to "hide" one's physical "flaws" by selecting the proper cut and color for one's "body type"), encourage viewers to become more selective in their consumption, in order to consume the product being offered. This is one of the main tenets of the pedagogy of consumerism elucidated on talk shows. As Ewen (1976) illustrates, people in the Fordist context needed to be educated on consuming in order to use the over-produced goods of industrialized America; consumer culture became a worldview which inculcated people into the habit of consumption as a path to a fulfilling life.

We argue that this process is continuously re-formulated to suit a particular stage or phase of capitalism. Indeed, in the twenty-first century, people in the techno-capitalist context still need to be reminded to consume, and to consume a particular brand, despite movements of ecological and environmental sustainability[10]; in fact, they need to re-negotiate their consumption to better adapt to the changing global situations. For example, the new liberal "green mainstream" appropriates the rhetoric of environmentalism yet does not aim to significantly change structures that produce environmental degradation, pesticides, and pseudo-organic plants. By featuring green stories and somewhat environmentally friendly products, talk shows offer a pedagogy of consumption that is flexible enough to justify and legitimate globalized consumer hegemony.

In this commercial venue, spectacle determines market value, as families appear on television to throw turkeys at one another, fight over their women or men, and reveal a shocking secret to their fiancé(e) before the wedding. In some types of shows, the personal (private) is made public through the use of confrontation, and individuals' problems, whether important (or valid) or not, are transformed into commodity.[11] The make-over is a classic talk show confrontation tactic to inculcate lower-income people into re-activating consumerism by giving them Gap or K-Mart clothes, a new hairstyle modeled on *Friends*, and new cosmetics from Revlon. Sometimes representatives from clothing companies and outlets, hair salons, make-up companies and fashion magazines such as *Vogue* parade audience members around the stage, discussing the "improvements" they have made to their style. Invariably, the changes involve becoming more middle class and mainstream. Types of make-overs include compelling rebellious young people to dress like "normal humans"; encouraging men to shave their beards, cut and style their hair, and wear jackets and ties; making larger women wear "flattering" clothing; and forcing mothers to dress "appropriately," or with longer skirts (processes to be addressed in chapters 5 and 6).

Even psychologists appearing on talk shows help work out such ill-defined "problems" as daughters who dress like vampires, mothers who dress like prostitutes, and large-breasted women who dress "too sexy." Usually the psychologist/therapist is introduced as an expert who just wrote a new self-help best seller. Often times, the therapist gives incomplete information and then concludes with, "the rest of the solution is in my book; you should buy it," to which Montel Williams advises "buy the book." Not surprisingly, every talk show host has written a book. Some are autobiographies, others have done cookbooks (Oprah Winfrey, Kathie Lee Gifford), workout books (Winfrey), books of children's art (Rosie O'Donnell), inspirational books (Williams, Gifford), revealing behind the scenes looks at the shows (Jerry Springer), and pop-psychology workbooks (Dr. Phil McGraw and his son Jay McGraw).

Through content commodification and intertextuality, talk shows' intertextual commodities help blur the boundaries between program and advertisement, as the ads become topics around which shows are structured, such as a celebrity interview with an artist who showcases and distributes to the audience their new album, such as *Oprah's* Celine Dion's "Miracle Babies" episode in January 2005. Dion sang songs from her new album; she distributed the album and the tie-in Anne Geddes & Celine Dion Miracle 2005 baby photography calendar. Sometimes an artist and album happen to be featured in an upcoming film, which is also plugged on the talk show. As a result, *The Oprah Winfrey Show* has become the most desirable advertising venue for musicians' new releases. Music industry executives have even admitted that the *Rosie O'Donnell Show's* and *The Oprah Winfrey Show's* ad-friendly atmosphere helps sell CDs and other products, and that this practice is changing the shape of daytime television. For example, Kenny G, Rod Stewart, Madonna, Clint Black and New Edition saw their record sales double in the week following their appearance on *Oprah*.[12]

Another example of content commodification is *Oprah's Book Club*. Every month since September 1996 (with a brief hiatus), Winfrey has announced a book that home viewers should buy and read. A month later, one taping of the show entails a select group who eat dinner with Winfrey; the dinner tape is later used as an intertextual device to be played during a studio discussion of the book. The purpose behind this project, she claims, is to get America reading again and to give attention to books that are life-changing. Each book for the *Book Club* has immediately skyrocketed to tops of bestseller lists[13]; Winfrey's staff now calls and alerts bookstores and libraries ahead of time, so they can place a larger book order to prepare for the crowds who will visit the mall to buy the latest *Oprah Book Club* book. Additionally, Starbucks Coffee sells the *Book Club's* books in their stores, in a Starbucks/Oprah literacy campaign. Winfrey makes no money from the sales, and Starbucks donates net profits of the books to a literacy training foundation within The Starbucks Foundation, actually keeping the profits in-house by allocating the money to another company division. In this way, talk shows support second-order products, which are sold on the programs. These acts promote conspicuous consumption.

Conspicuous consumption on talk shows has been helped immensely by Rosie O'Donnell. Before a spattering of negative publicity (unfairly launched at her coming out as a lesbian), and a lawsuit against her and her magazine, and her quitting the show, which hired new host Caroline Rhea, O'Donnell was deemed by market gurus, the "Queen of Nice." Her talk show was conceived of and billed as an advertiser's dream. At the start of each celebrity-driven interview show, she would announce that every member of the studio audience could find a Ding Dong and a Koosh Ball under her/his chair. She also reserved the first segment to demonstrate all of the promotional products she received that day, from M&Ms to plastic horses, to Twinkies, to CDs, Barney dolls and Disney figurines, Warner Brothers' stuffed animals, and Nickelodeon's Rugrats videos. Sometimes, O'Donnell programmed special promotional weeks centered on a product. Once, she featured "Thong Week," during which she tested five companies' thong underwear. Each day she sang an original song about thongs, revealed which brand of thong she was wearing that day, and inspired the audience members to wave around the free thongs located under their chairs. Another intertextual delight involved a news story of two six-year-old white girls who had set up a lemonade stand on their street to raise money to buy new adidas sneakers for the school year. Unfortunately, someone ran by, stole their money and demolished their stand. To an audience of oooh-ing and aaah-ing adults, the girls told their heart-breaking story. Afterwards, a corporate-speak Adidas representative walked onstage and awarded the girls a prize: he and Adidas would "outfit" them in Adidas shoes, clothes, and accessories until they graduated high school. O'Donnell led the audience in their praise of Adidas' generosity.

Both O'Donnell and Winfrey also launched their own magazines. "O" and "Rosie" provided a perfect tie-in to the programs. *Oprah* often shows the magazine on screen and bases the episode on the articles in the magazine. Her "Change Your Life" season consisted of special guests to be featured all season on the show and to write a column in the magazine. Gary Zukov, Iyanla Vanzant, (both to be discussed further in the next chapter) Suze Orman, and Phil McGraw each with their areas of expertise, doubled as content for both television and magazine. These characters were so well utilized and branded that they each have their own books, and several had or have their own talk shows (Vanzant, Orman, and McGraw). In this way, talk show spin-off commodities have homogenized the television and magazine scene. Rather than take a risk on a new, unique talk show character, the proven success of Orman's "O" and *Oprah* spots induced MSNBC to offer her a financial planning program. Already a branded commodity with a loyal *Oprah* following, Orman's program speaks to the upper-class viewing audience: one can become successful at saving money, she advises, by skipping the daily $5 latte. Likewise, Dr. Phil McGraw, already blessed by *Oprah's* validation, has seen multiple book contracts for himself and his son Jay McGraw; the show is based on the themes in his books, and guests are requested to perform his workbook activities for self-help homework. His branded image was a safe bet for advertisers and book publishers alike.

Winfrey herself spun off two shows onto the recently formed Oxygen television network. The first program revolved around Winfrey being taught how to use the Internet, a tie-in to Oxygen's Website. The second program, *Oprah After the Show*, is a mere continuation of *The Oprah Winfrey Show*. It is produced directly following her original program, but guests and audience members stay "after the show" taping to tape further. At this point, the atmosphere becomes more relaxed and informal, and audience questions are sometimes taken. In this respect, *After the Show* harkens back to earlier *Oprah* days of audience participation. This spin-off is low cost, as the entire production is merely a half-hour extension of the more heavily produced *Oprah*. The fact that the show is on Oxygen is no coincidence, however; Winfrey was one of the original handful of private owners of the network. Thus, she owns a stake in the profits of the company, as well as her program.

As evidenced, talk shows socialize viewers by appealing to a sense of reality and practical information. Ironically, though, talk shows' sense of reality is actually one of hyper-reality.[14] Constructed stories, produced for ratings and profit, spreading a pedagogy of consumerism and celebrity worship, are naturalized, as evidenced in the theme songs of the programs. For instance *Leeza's* theme song was: "this could be the best time of your life"; *Jenny Jones*: "So real, everyday people. Make it real"; *Oprah Winfrey*: "get some fire, get inspired, get with the program, Oprah." These attempts to construct the programs as "real life" shape what the popular imagination idealizes as real/normal—these constructions of reality are deeply rooted in classed, gendered, and raced assumptions, to which we return more fully later in the book.

Content commodification thus produces "creative, economic, and cultural implications."[15] The creative process is transformed, making room for advertisers earlier in the writing and producing process and opening the door to further commodification. The talk show genre thus becomes more formulaic and restricted, as their reliance on revenues from intertexts limits the types of shows they produce.[16] Shows investigating abuses in the American health care system, genocide in Sudan, or the Bush administration's war in Iraq are unlikely to provide easy tie-ins to commercial products or excited advertisers of multinational corporations. Thus, vulture culture is perpetrated on talk shows through their commodification of social problems and de-politicization of culture.

COMMODIFICATION OF THE AUDIENCE

Let us turn, then, to the second area of commodification. Talk show audiences play an important role as media institutions bind together the audience, media, and advertisers.[17] The reciprocal relationships are such that media "sell audiences which perform three key functions for the survival of the system: audiences market goods to themselves, they learn to vote for candidates in the political sphere, and they reaffirm belief in the legitimacy of the politico economic system."[18] Essentially, advertisers buy access to audiences. Thus we witness the proliferation and concern over audience demographics and marketing shows to specific demographic groups, such as women aged 18 to 49. Since shows are sold to niche markets, advertisers can better spend their money by directly targeting desired audiences. As such, talk shows become vehicles for slicker marketing ploys and ratings races.

In order to obtain higher ratings and advertising dollars, talk shows need to make themselves suitable for advertisers, as links to the commodified audience must be made solid. As seen in the previous section discussing commodification of content, talk shows become advertisements for second- and third-order products through intertextual tie-ins and special feature segments involving a marketed product, especially other media products. With "advertisers' revenues setting the context within which popular culture production takes place,"[19] advertising potential directly affects content and a show's particular hailing of the audience.

Oprah Winfrey's popularity directly reflects this process of attempting to gain audience share. Although Winfrey is a popular celebrity, her show is critically acclaimed and has won numerous awards, the show's ratings, thus potential for ad revenue, waned somewhat in 1994 and 1995. Some critics claim that this is due to newcomers such as *Ricki Lake* grabbing ratings from younger demographic groups. Another factor was that Winfrey's "decision to take a higher road in the subjects covered, at a time when TV talk [had] become increasingly criticized for pandering to the worst instincts of the general public,"[20] prevented the show from successfully competing with the continuing spectacle on other programs. Controversies over subject matter and production tactics escalated until the tragedy befalling guests of the *Jenny Jones Show* in March 1995. Scott Amedure came to the show concerning secret crushes to announce his interest in his friend, Jonathan Schmitz. The homophobic Schmitz became outraged at his "ambush" on national television, and shot and killed his friend three days later.

Shortly thereafter, in October 1995, a conservative coalition, Empower America, led by William Bennett and two Democratic Senators, Joseph Lieberman (CT) and Sam Nunn (GA) joined together to denounce talk shows for the "cultural rot" they were promoting. They planned an attack on all shows not conforming to their standards. Lieberman claimed that this was not an act of censorship, but "'an effort to confront the people at these companies and get them to consider what is coming out of the end of the pipeline they own.'"[21] Enlisting the help of companies frequently advertising on daytime talk shows, advertisers were asked to boycott the shows until "the sleaze mongers...clean[ed] up their act or [got] off the air."[22] They used the argument that the shows are "harmful to children," even though in reality, this demographic only constitutes about 6% of talk show audiences.[23] Corporations were urged to restrict advertising funding towards certain types of content. Kelloggs, AT&T and others pressured the shows to respond. Proctor & Gamble even pulled ads from seven talk shows.[24] The programming context of this time period was that the O.J. Simpson trial coverage and proliferation of new talk show clones depressed ratings that year; why would advertisers support shows whose ratings and shares were slipping, and therefore not reaching as large an audience for their dollar?[25] The proliferation of media products moves "advertisers to the most effective medium as the cost of each becomes uneconomic."[26] Consequently, the top of the market benefits from ad money, while the bottom consists of struggles for larger markets and economies of scale which result in a homogenization of the talk show content.[27] The ultimate commodification of the audience not as concerned citizens but as a buying public is exemplified in Empower America's coalition with advertisers and their "concern" for the audience's moral sensibilities. The audience, their ratings, as a link between programs and advertisers was exploited in order to nudge shows to make content decisions to please a political organization, the advertisers, and the audience only as a function of the advertisers.

Before Empower America named names, Winfrey decided to change her format, to prevent being singled out by the coalition but also to combat her slipping ratings, and was applauded by the group for sanitizing her show.[28] This "sanitization," then, can be conceived of as a political economic rather than purely aesthetic and moral move. Her show continues to focus more on practical issues and topical narratives than previously. In fact, her commodified spectacle continues to amaze, as she presented cars to an entire studio audience at the beginning of the 2004–2005 season. This type of self- regulation in the face of public disapproval is typical whenever business is threatened. In one survey of general managers, 37% said "they have canceled a talk show or considered doing so because of the content,"[29] and 43% said "that talk show [advertising] spots are more difficult to sell because of concerns over content."[30] Concern for profit drives the advertisers and producers, which then drives content and its regulation, in order to remain suitable for advertisers.

Empower America's witch-hunt is reminiscent of the Red Scare in the days of McCarthyism. This movement impacted the talk show genre in the 1950s, playing a central role in the development of the "chatty" prong of talk shows, rather than allowing the serious issues-oriented programming criticizing big business as corporations and government tried to bust labor unions. The 1990s backlash used big business once again to reign in so-called deviance under the facade of getting to the bottom of the problem, the "cultural rot" of talk shows themselves. Ignoring the political, social, and economic conditions under which the shows are produced and aired, and those within/against which the guests are trying to live their lives, the gatekeepers tried to "clean up" the airwaves. Winfrey played into the criticism by making content changes that greatly affected the production, content, and consumption of her show, by commodifying her program further into the land of programs having a "suitable atmosphere for advertising" so that her show could regain its money-making status.

The link between so-called questionable content and being difficult to sell is the link that connects political economic influence to content control. During Empower America's crusade, development for the *Rosie O'Donnell Show* was already underway, to be released in June 1996. From its conception, the show was billed as "advertising friendly" and was the only show picked up in 53% of the market in a year when talk shows were the dregs of syndication. As a syndication sale benchmark, in 1995, "Carnie," a quickly canceled program, opened up to 70% of the market—a considerably larger market than the fanatically successful *Rosie O'Donnell Show* sold one year later.[31] Thus was born *Rosie O'Donnell Show*, the most advertising friendly talk show to exist since the days of direct corporate sponsorship (she has featured praise for Listerine, Adidas, Victoria's Secret, M&M, Microsoft, GM, Barney, and many, many more). The threat of competition her new program posed, we argue, constituted another factor in Oprah Winfrey's decision to revamp her own programming—if the trend was shifting, she was sure to be at the forefront and continue to profit.

In the months following the "Jenny Jones murder" and Empower America's assault, talk show ratings tapered off. However, after the initial hype, and the canceling of excess shows which had oversaturated the talk show market and thus made advertising scarce, the ratings began to rise. In 1997, *Oprah* was number one in the early fringe; *Rosie* had shot through the roof, becoming the top rated show in her time slot; and *Leeza* was up 49%.[32] Before 1995, *Oprah's* content resembled other shows, including *Jenny Jones, Jerry Springer, Rolanda, Ricki Lake, Montel Williams, Leeza*, and *Donahue*. This type of diverse programming—part entertainment, some journalistic stories, a few celebrity stories, and much spectacle—would now probably be considered "trash" and therefore a risk for advertisers who do not want to offend righteous viewers with spending money, or members with potential macro-level power. The show would focus on a topic, showcase tell-all guests who would appear for the duration of the program, and who would be assisted by an expert who would help resolve their problem. Alternately, they would either conduct a celebrity interview or cover a "newsworthy" story.

Oprah still features interviews and stories, but the format of the show is broken up into smaller, more malleable segments. Her 1999 season titled, "Change Your Life TV," aimed to address "important" issues in the face of talk show criticism. Her new set was cozy, decorated in warm colors, plush chairs, and filled with large screens and more technologically advanced televisual effects during special segments. Winfrey has shifted from highlighting controversial and confrontational guests toward a more middle-class/elite discussion-oriented focus on "practical" issues, such as home decorating and baking tips, child care issues, and female bonding activities (such as audience-wide make-overs, fashion shows, and the *Oprah Book Club*). Perhaps more interesting is her overarching focus on philanthropy, volunteerism, and spiritual well-being, three issues which are brought out in special segments and projects with which Winfrey is involved. All of these issues tie-in commodities in other sectors, and the talk show as intertextual advertisement increases the surplus value of commodities in other sectors of the economy, as in the CD and *Book Club* examples.

One final example involves the creation of the *Oprah Angel Network* (riding the high tide of angel rhetoric sweeping the country) which encourages volunteerism and philanthropy at both the individual and corporate level. She has initiated the World's Largest Piggy Bank collection, only allowing coins to be deposited, the proceeds of which were donated to minority college scholarships Winfrey created. Corporate sponsors can donate money to sponsor a house-building project, in which Winfrey has aligned with Habitat for Humanity to sponsor "200 houses throughout the United States—one for every Oprah television market."[33] Companies such as Merrill Lynch, American Airlines, Owens Corning, Sargento Foods, and Mutual of Omaha Companies donated money for one house each, and have appeared on the show to receive a trophy, a handshake, and praise from Oprah Winfrey. These two projects illustrate liberal tendencies to attempt to fix problems within the existing structures without trying to change the structures. This is the difference between a radical act and a liberal one, the difference between rewarding these companies and Winfrey for promoting a so-called responsible capitalism and acknowledging that no capitalism is responsible, for it is inherently exploitative. While it is commendable that Winfrey herself donates millions of dollars to charity, it is only because she has earned and continues to earn hundreds of millions. Her generosity is, once again, kind, yet we need to question its efficacy in changing the root causes of socially unjust institutions. Rewarding multi-million dollar conglomerates for donating tax-deductible pocket change in exchange for a free advertisement and the approval of Oprah Winfrey is an extremely limited form of social action. Corporations actually come away from such an experience looking like model citizens—but have they changed their pay scales such that the janitors at the company have the same benefits as the CEO? By promoting the ideology of philanthropy as a social cure, the audience is commodified as a niche market of "responsible consumers," a more middle-class market that can be targeted more successfully with ploys of upward mobility through consuming the products of philanthropically generous companies. Thus, despite her attempts at "changing lives," it is possible to say that Oprah Winfrey falls short in creating or promoting a radically just society.

COMMODIFICATION OF LABOR

Another aspect of commodification obscured by the dazzling lights of the talk show is related to those who turn the lights on and off. The labor that goes into researching, writing, and producing talk shows is important in understanding how talk shows play a role in media culture. Labor, in all capitalist institutions and organizations, is exploited in order to gain a surplus value for the owners of the capitalist class. Labor exploitation is traditionally of two types: absolute (meaning laborers work longer hours for the same or less pay), or relative (labor is intensified—workers complete more work in the same or less amount of time). Talk shows exhibit both types of labor exploitation in their use of camera operators, script writers, producers, costumers, interviewers, screeners, and other workers who do not make as much money as the celebrity host, production company, or distribution company. However, talk shows perform a new and improved exploitation as they "employ" an entire studio audience of unpaid volunteers, as well as a panel of guests who appear for free. In this way, talk shows play on people's desires to be famous and appear on television, as a way to exploit labor at no cost—the shows purely profit. Many times the guests have their hotel and travel costs covered by companies donating their services to the show. These services are exchanged for a promotional credit at the end of the show, in essence a short advertisement, sometimes read by the celebrity host. For example, at the conclusion of *Sally Jesse Raphael*, an announcer would read text appearing on the screen: "Some guests in our audience receive and a promotional fee has been provided by..." Oprah Winfrey thanks American Airlines for providing transportation for her guests. Not only is this commodification of labor, but again, an advertisement for the companies that have donated products or services.

Finding guests to appear on the shows constitutes a task that comprises some workers' entire job descriptions. These guests can be courted by announcing a topic and phone number at the end of the program, and asking potential guests to call the show if they have that problem and would like to appear to discuss it. When guests call in, their names and problems are recorded by a production assistant, who then passes on "interesting" situations to interviewers who may call the guests back to discuss the possibility of appearing on the program. This phone work saves time when the guests arrive, as they have already been pre-screened and interviewed. Some shows, such as *Montel Williams*, *The Oprah Winfrey Show*, and *Dr. Phil* have developed a following and created relationships with viewers who feel compelled to write unsolicited letters to the hosts themselves, in the hopes that they can appear on the show and discuss their issues. The names of guests that have indeed appeared on shows are collected and put into a guest directory.

Directories also exist for experts. However, if a show does not want to sift through the expert directories, they are not at a loss: experts who have just written books—or their managers/publishers—contact shows to solicit exposure. Likewise, the celebrity solicitation of musicians, actors, and politicians to appear on talk shows such as *Regis & Kelly*, *The View*, or *Ellen DeGeneres* removes added labor of finding guests. As these people vie for interviews, again, a free outlet for celebrity commodification overlaps with the commodification of labor. As a concrete example, *Leeza*, an NBC-owned show, would commonly feature actors from other NBC programs, mostly soap operas (*Days of Our Lives* and short-lived *Sunset Beach*), as guests. This tactic of using monopoly control provided intertextual advertising for other NBC programs, as well as attaining free labor by having their own actors appear as publicity stints to support other NBC programs. Finally, they did not have to call, write, fly or otherwise coax guests to appear—these guests pop over from one studio to the next on their lunch break to show their faces and mention the name of their own program.

Additionally, one special theme often used on talk shows is that of the "update." This tactic involves the host updating the audience and viewers on former guests who were of particular interest. Updates involve showing clips from the original program, and either inviting the guests back to check on their progress, or else an announcer or host simply reads a summary of a phone interview conducted with the former guest. For these programs, barely any production costs are spent, in that most or all of the show's content is simply recycled from earlier programs. At most, a writer and editor must sift through the tapes to find suitable clips.

In conclusion, commodification and commercialization are cornerstones of vulture culture. Talk shows' participation in tie-ins, merchandising, and celebrity features shapes daytime television's pedagogy of consumerism. Consumer culture as a cultural ideal is produced, reproduced, and repackaged for contemporary talk show viewers, who are themselves commodified in their roles as advertising viewers and ratings producers. The commodification of talk show labor also places talk shows in a cultural category that harkens to wider social processes of downsizing and temporary/transient staffing. Vulture culture feeds on these commodifications of cultural, social, and political life. In the scavenging of content, audience, and labor, talk shows privilege profit imperatives over the guests, problems, and solutions they portend to define.

4

TALKING SENSE:

COMMON SENSE AND EXPERT KNOWLEDGE

When Dr. Phil McGraw's daytime television talk show debuted in 2002, the following terms were used to describe it: it "is a tough-talking Texan who dispenses no-nonsense relationship advice."[1] Dr. Phil's Web site reminds us that on his debut season, he "garnered the highest ratings of any new syndicated show since the launch of *The Oprah Winfrey Show* 16 years prior." This is so because "*Dr. Phil* combined enlightenment with entertainment value as the show took on such topics ranging from human functioning to behavioral medicine to legal issues." Other advertising statements put the same emphasis on what is perceived to be one of McGraw's major qualifications for his show's anticipated success: his ability to manipulate both enlightenment (expertise) and commonsense knowledge, his tell-it-like-it-is style.

A close analysis of the evolution and development of daytime talk shows reveals that expertise and common sense have been part of the secret of these programs' success and large public appeal. Thus, McGraw is not bringing to the talk show a completely new approach or vision about the type of knowledge most suited to the show's format. If anything, he reassures his potential public that he will remain faithful to that which has always attracted, fascinated, entertained, and even angered them: common sense and expert talk. According to his own Web site: "Dr. Phil McGraw has single-handedly galvanized millions of people to "get real" about their own behavior and create more positive lives."[2]

This chapter demonstrates specific instances in which vulture culture appropriates, repackages, and presents various forms of knowledges, expertise, and experiences. Common sense and expertise are particularly important to explore in relation to vulture culture because it is precisely through the intersection of common sense and expertise that people make sense of their everyday lives. Understanding the ways in which vulture culture recycles information and knowledge is crucial to a critical pedagogy and media literacy.

A critical pedagogy of talk shows, as a manifestation of vulture culture, should address questions related to knowledge production, information circulation, legitimation as well as the power and processes of naming and/or authorizing information.

This chapter discusses the complex and intricate relationships between identity, power and knowledge as they are played out in the talk shows' dynamics. The first section, then, examines the category of "expertise" in tandem with that of "common sense" in order to underline the participation of the talk show culture in, and its effects on, the processes of knowledge production. Discussion of the expert/commonsense knowledge is important here since it allows a better understanding of the multiple roles taken by the programs' hosts, guests, "professional" speakers, as well as studio audiences.

The next section of the chapter analyzes specific examples of hosts' individual biographies in order to reveal the weaving of the biographical into the professional—a process that confers on hosts a significant power to judge, admonish, punish, and/or reward his/her show participants and viewers. Although the chapter is not centered exclusively on the role of show hosts, the questions raised here invite a closer investigation of the power created by and conferred upon hosts through their celebrity status.

DEFINING COMMONSENSE KNOWLEDGE

The term "common sense" is used in phenomenological sociology to refer to the knowledge drawn from the fundamental reality within which people live, and interact with others as well as with their larger environment.[3] Such knowledge is "routinely" relied on in the conduct of everyday life. Researchers' interest in the nature and functions of common sense has added useful clarifications to the term that are pertinent to our understanding of the types of knowledge circulating in the talk shows arena.

According to Antonio Gramsci common sense is "the philosophy of non-philosophers": it is the experimental, practical, fragmented and contradictory knowledge of the masses. It is, in other words, "the conception of the world which is uncritically absorbed by various social and cultural environments in which the moral individuality of the average [person] is developed."[4] The function of this knowledge in society is important to underline: common sense works as a "glue" that gives unity and legitimacy to the dominant views, institutions, and centers of powers in any given society.

In essence, this knowledge is neither innocent nor is it neutral. It works towards maintaining the status quo and presenting it as natural, hence acceptable to the majority of people.[5] Common sense naturalizes dominant discourses and views while presenting them as legitimate references or sources of knowing.

To this type of knowledge Gramsci juxtaposes the more theoretical and potentially critical knowledge of the "elite" who transcend the incoherences and fragmentations of common sense. It would be misleading, however, to assume that common sense stands as the opposite of expert knowledge. The relationship between these "knowledges" is much more complex and less clearly defined than the terms of opposition might suggest. Before we discuss the complexity of this relationship, a few comments on expertise are needed here.

The astounding number of existing "experts" is one of the defining characteristics of our technologically advanced, consumer-oriented and anxiety-laden societies. Anthony Giddens, who has done extensive work in this area, maintains that modern societies are thoroughly organized and guided by expert systems, so much so that our daily life is placed under "the auspices of technical and professional knowledge."[6]

Thus, whether we are talking about fashion wearing, body shaping, relationship building, home managing, teenager controlling or addiction curbing, we have a group of experts who promise us practical advice or a course of action that is based on "scientific" findings. To virtually each and every aspect or experience in our lives, there is a "matching" profile of expertise that claims understanding, moral support, and practical assistance. This wealth of expert knowledge and advice finds expression in the considerable amount of books, manuals, magazines, and videos that offer "ten," "twelve," or "fifty" ways to solve this and that conflict or attain this or that objective in life. The phenomenon is impressive both at the level of the quantity as well as varying quality of the knowledge produced. These products are becoming a familiar sight not only in bookstores and libraries, but also and perhaps most imposingly now on the stages of daytime talk shows. Can we imagine shows like *Montel*, *Oprah*, *Sally*, or *Dr. Phil* without the voice of expertise in one form or another?!

True, not all daytime shows rely on experts with the same "religiosity" or regularity. *Jerry Springer*, for instance, moved away from such a tradition long ago. Yet, the important question that begs analysis here is not whether daytime talk shows may or may not exist without different voices and expressions of expertise. The questions are related to the specific use(s) of expertise on the show, the relation of expertise to common sense, and the meaning of these knowledges with respect to issues of power and authority.

Michel Foucault has given us ample examples in his work about how expert knowledge in any area of experience confers a considerable amount of power on the experts and on the social order that supports them. This insight is most helpful in understanding the complex relations between power and knowledge on the one hand, and the subtle nature of power inherent in the language of professional expertise on the other. Taking the specific example of psychiatry, Foucault convincingly demonstrates that the "psychiatrization" of everyday life, if it were closely examined, might reveal the invisible hand of power."[7] If we closely examine the kind of discourse promoted by either the talk show experts or the show hosts themselves, then one begins to understand what is meant by the "invisible hand of power."

HIERARCHIES OF KNOWLEDGE/HIERARCHIES OF POWER

Daytime talk shows, as this book seeks to demonstrate, are sites of knowledge production and consumption where the diversity of competing discourses both reflect and shape the construction of cultural practices and the formation of identity in contemporary American society. These programs regularly stage knowledges that may be complementary at times and conflictual at others, and as such, they often problematize questions of truth, credibility, and power through the complex relationship between "expert" knowledge and "common sense."

It is important to remember that daytime talk shows borrow from a diverse set of knowledge disciplines and discourses as well as a constellation of voices. In terms of knowledges and discourses, the shows situate themselves between the confessional genre, the investigative reporting, unsolved mysteries, autobiographies, psychotherapy sessions, and religious sermon.

As far as the voices are concerned, the ones regularly present on the shows are those of the lay people, members of cultural and socio-economic marginalized groups whose personal narratives and experiences give these programs the primary reasons and substance to exist and flourish. Equally important are the voices of experts— doctors, psycho-therapists, psychiatrists, social workers, financial advisors, family counselors, etc.—representatives of the professional and technical spheres of knowledge. A later section of this chapter will reflect both on the importance of the show host voice and the studio audiences, as well as ways in which guests contextualize the expert's credibility.

This brief reminder of the multiplicity of voices and knowledge sources on the shows aims at underlining the fact that the talk show format lends itself to the construction of authority and the reproduction of relations of power on these programs. This is so because while the shows' participants are invited for the pertinence of their life stories and experiences, the experts are appealed to for the relevance of their professional know-how and their "technical" reflections on personal narratives. To a certain extent, the experts enjoy the privilege of commenting on other people's private lives and choices precisely because the label of "expertise" gives them the power to do so. The guest speakers, on the other hand, might not agree with the expert's views or analysis, but they do not question the expert's status of power itself. This is a subtle but important distinction to make. Carpignano et al. offer an interesting insight into common sense and the talk show, one that will resonate throughout the rest of this chapter.

The talk show rejects the arrogance of a discourse that defines itself on the basis of its difference from common sense. On the other hand, the common sense of the ordinary public is not at all based on a set of natural intuitions, consolidated with the life experiences that are brought to be into balance with that of experts. Their common sense shows its conventionality, its social training, its knowledge of television conventions of speech and gestures.

It is important to restate that it is misleading to assume that the shows' different guests are bearers of clearly defined categories of knowledge that stand in opposition to each other: lay people/commonsense versus experts/professional knowledge. What needs to be stressed is the interpenetration as well as fusion between the knowledge sources of the shows' guests, audiences, and experts. This interpenetration takes place in various ways.

First, although "common sense" has been referred to as the philosophy of ordinary people, such a category of knowledge is not produced out of a vacuum. Rather, it is constructed out of people's continuous exposure to political, medical, legal, economic, and scientific explanations, among other discourse.[8] This is precisely what we mean by vulture culture: this process in popular culture of picking and borrowing from as well as fusing different areas of knowledges, experiences, and discourses.

The exposure to this vast corpus of knowledge takes place on a regular basis whether through the different media channels and products or directly through personal experiences with paid experts. Alfred Schutz's phenomenological reflections on the "social distribution of knowledge" are pertinent here. "As a matter of fact" he states "each of us in daily life is at any moment simultaneously expert, well-informed citizen, and man on the street, but in each case with respect to different provinces of knowledge."[9]

In other words, lay knowledge is neither pure nor uncontaminated by the language and logic of the expertise that is abundantly available in our contemporary information landscape. Thus, personal guests who bring to the talk show their personal narratives are not totally unaware or incognizant of the plethora of expert reflections and advice. If anything, the show guests also bring with them a sizable capital of "expertise" they have accumulated over the years and have invariably interacted with prior to their appearance on the show. Also, the show guests are often regular viewers of these daily programs, hence, regular consumers of the expertise circulating on the shows.

On the other hand, our information society is such that various types of expertise and technical knowledge are becoming readily and increasingly accessible to almost all. This type of knowledge constantly flows through different channels of communications, in various media forms and genres, among which talk show programs and their media products play a major role. Transmitted through different media of information and communication, expert knowledge no longer retains its "esoteric" or inhibiting aura of technicality: it presents itself in the simplified language most accessible to the ordinary people.[10] It is in this simplified form that expert knowledge reaches daytime talk shows and it is this accessible language that produces the diverse products promoted on the daily shows. Vulture culture in fact thrives on simplicity and simplifications. It makes it easier for the viewer to consume a multitude of ideas and discourses without seeking to trace their origin or authenticity.

Talk shows not only tease us daily with the question of how common is "common sense," but they also problematize the relationship between common sense and expertise. The two categories of knowledge thus feed into each other, while the legitimacy and power of each keep shifting in the discursive spaces of the shows. Put differently, it is misleading to believe that talk shows automatically favor the "scientific" voice of the expert over the experiential knowledge of the lay person.

There certainly is no ipso facto equation here. If anything, talk shows' commercial imperatives urge them to juxtapose the language of expertise and common sense, and create an environment for a spectacular confrontation between them. What is meant by this is that, every participant or guest of the show is a potential expert or can become one in his or her area of experience. Different voices get the opportunity to give an individualized version of "truth" or perception of reality while the hierarchical distribution of knowledge is momentarily suspended between the expert and laity.[11]

In doing this, talk shows seem to destabilize the opposition between "scientific" knowledge and the participants' "subjective" narratives that are informed by life experiences. Close attention to a large selection of the daily programs confirms that the expert's voice on various daily show is often challenged by both the programs' audiences and hosts while personal narratives are often prioritized on the basis of their perceived credibility and authenticity.[12]

Ethnographic evidence from a focus group organized around talk show viewing supports the above statement. As some viewers clearly put it:

> Erin: "I think its to make [experts] seem like they're trying to help people, but I really don't think they're actually trying to help people. Because, otherwise, they would set up counseling sessions during the show instead of just spending two minutes with an expert at the end of an hour-long show. I think they're very fake."
>
> Louise: "Yeah, I love it on *Sally Jesse Raphael*. I think it's Dr. Jean Sarillo. That I remember this is sad. [all Laugh]. Dr. Jean Sarillo is on every single episode. She's so so nonsense. She's like 'I talked to her and blah, blah, blah!' She just yells at people. She's like the boot camp guy too. She yells at people along with him. She talk to people for 3 minutes, figures out their whole life, tells them what to do and pats them off the stage, 'You need to get away from him.' It just seem so fake, you know what I mean?"
>
> Erin: "Yeah, and I usually disagree with what they say too because they do make a judgment and like, the role of s counselor or somebody who is trying to help people is to not say, 'You're wrong. This person is right.' It's to be like 'How can you improve it? How are you going to improve your life?' Not like 'You're wrong. You suck.' You know, I just think they are always wrong [laughs] basically."
>
> Rachel: "That's one good thing about Oprah does though. When she does have …"
>
> Louise: "the only thing about Oprah's experts is that I remember reading an article. Look, I've noticed this. Like Oprah's 'experts,' I put that in quotation marks because these are just people who write books, and Oprah reads them and is like 'Oh, that's good.' And then she has them on her show like they're experts, you know? And, I don't know, I think that's really sketchy just because she does wield a lot of power. And she holds these people up like 'Oh yes, they're expert, they know this and I'm Oprah.' You know what I mean? These people aren't experts, they're just, they got a book deal...."
>
> Rachel: "I don't think you can even call them experts. They're just authors [laughs]. They're just additional guests. They're just there to almost legitimize the show. Like we make them fight on stage, but we're trying to help them somehow. Like Erin was saying earlier they try to

calm things down towards the end, just like slow everything down before they little moment where they talk to the camera and things like that."[13]

The above discussion reveals that these viewers are struggling with the moral dimension and legitimacy of expertise on the show. Many of the above statements suggest that they do not feel comfortable with the idea of having an expert provide both judgment and guidance. Hence, they challenge the very idea what qualifies experts.

Also, the presence of the studio audience is not an insignificant feature of the shows' format. The audience is called upon, and is expected to act, as a representative body of "common sense", moral values and standards of behavior. And it is in these terms that it will re/act. In its own way, both the studio audience and the home viewers could be seen as the "expert" of common sense, and as such they display yet another dimension of expertise we should not overlook. The audience presents itself as the moral judge and witness of events in order to participate momentarily in the game of sanction, critique or approval of the guests' behaviors and opinions. There has been a gradual shift, over the last years, in the degree to which such hosts as Povich, Williams, and Winfrey solicit the comments of the studio audience. Generally, however, the audience plays a considerable role in determining the value, if not legitimacy, of the knowledges circulating on the shows. As stated clearly by a researcher in audience analysis:

> How audiences relate to a specific genre, what they consider to be of value and how they position themselves in relation to it all frame what they may gain from it, in terms of 'what knowledge' is at stake. If experts are considered to be lacking in personal experience while ordinary people are seen as authentic, the value of what each says will be regarded differently than it will be by those who consider that experts are more credible and more knowledgeable than ordinary people.[14]

Are we to understand, then, that the presence of experts on talk shows and the regular appeal to their social/scientific knowledge, regardless of the form in which it presents itself, are meaningless routines of the programs? Are we to suppose that audiences, both studio and home, are empowered interpretive communities who question the authorial voice of expertise? And, are we to assume that since the discourse of experts already informs the lay person's knowledge there are no struggles over power and authority in these daily programs?

It is hardly possible to answer any of the above questions in the affirmative. The staging of power relations on these shows cannot be overlooked if we take into consideration the central place occupied by personal "confessions" and therapy, as the privileged area of expertise. Daytime talk shows might be staging a confrontation between experts and lay people, but the spectacle they offer is not a celebration of any behavior that falls outside the acceptable dictates of society. The daily programs remain fairly constant in privileging conformity to dominant social values and norms. The programs are about the exposure of what society considers "anomalies" that require a return to "normalcy" with the help of professional guidance. Experts might then be resented or challenged, but the shows clearly present them as the voices of wisdom and sanity.

Confide in Me, I'll Confine You

There is little doubt that what makes daytime talk shows mostly compelling to some and repelling to others is their obsession with the emotional, psychological, and physical dimensions of individuals' experiences. The confessional and therapeutic aspects of the shows have attracted the attention of some researchers who have tried to establish linkages between the nature of these aspects and the audiences they are targeting.[15]

Thus, it is established that since women represent about two thirds of the viewing audiences, the programs extensively draw on the earlier journalistic tradition of providing women with advice, and specifically of a therapeutic kind.[16] This is still the case with the most recent additions to the list of talk shows, such as *Dr. Phil* despite the fact that programs such as *Springer*'s or *Ricki Lake*, target a wider circle of audience viewers across gender, class, race, and age.

Women thus remain heavily targeted by the talk shows and the related self-help products they promote. Among the most popular concepts on the programs, and the ones most frequently used by experts, hosts, and audiences alike, is the "empowerment" of women through a greater degree of self-reliance.

In 1999, *The Oprah Winfrey Show* introduced a series of shows titled "Change Your Life." Here Winfrey featured a group of guests/ experts whose talents and knowledge would help better the lives of millions of viewers. An extended example from an appearance by Gary Zukav on Oprah's "Change Your Life" series makes this point clearer. Gary Zukav speaks about a concept he calls "authentic power." In this episode, Zukav manages to naturalize individual change and reconciliation through the liberal, therapeutic, and generic religious individualisms.[17] He starts by saying that to understand authentic power, you must first understand its opposite—which he calls external power. According to Zukav, external power can be anything from clothes to addiction to a man or a job. In contrast, authentic power comes from inside and can never be taken away (liberal individualism). The program continues to discuss the lives of formerly wealthy people who have re-assessed their "external power." Initially, it seems as if authentic power might offer a critique of capitalism. However, authentic power's focus is once again on the individual. It becomes a way of defining yourself and being aware of your surroundings. As Oprah suggests, "I have a lot of things, but I am not defined by any of those things."

Authentic power does question wealth, but not critically. Zukav suggests wealth is an external power. But, you can be wealthy and authentically powered if you have the (amazingly vague) four characteristics of an authentically powered person (therapeutic individualism). The first one is humility where you realize everyone's life is as difficult as yours. The second one is forgiveness, without which you see the world as a dark place. The third is clarity which says your experience is tailor-suited for you. Thus, you are not a victim regardless of race, class, or gender (As we will see, this is similar to another *Oprah* expert Iyanla Vanzant's "Wherever you are in life is where you're supposed to be.") And finally, an authentically powered person lives in love. This basically means you have empathy. Zukav gives the example that you realize that every person who cries hurts.

In this way, authentic power is an individual power with an emphasis on changing the self and coming to an acceptance of the world as it is. Accepting your current circumstance, authentically powered persons can thrive in any environment so long as they see everything that comes to them as a gift. In this way, all of our lived experiences of injustice and inequality are spiritual challenges we learn and grow from, and thus we have a merger between the discourses of religious and therapeutic individualism. Put in this context, the fact that we can challenge and change the socioeconomic conditions that create injustice becomes irrelevant and unnecessary. This episode on authentic power never challenges the factors that create the desire for wealth nor is there a critique of capitalism. Instead, the individual is told how to change and the pleasure of consumption is validated.

However, the logic of empowerment and the language of "coping" that usually circulate on the programs carry with them a strong residual dimension of more self-discipline and conformity. In other words, talk shows might give voice to the culturally and socially excluded working-class black mother, and might valorize the story of the racially or ethnically marginalized youngster, but they do so according to a vision that recreates as well as reinforces the logic of self-help movements. Issues are individualized, personalized, and detached from the larger sociocultural and economic environment in which they occur. Likewise, crises seem to begin and end with the individual: their origin as well as solutions revolve around the atomized "self."

In order to better understand this argument, as we believe it will be key in understanding common sense and talk shows, we will briefly delineate how Western thought defines and constructs the individual. The logic of liberalism is characterized by a belief in progress and the autonomy of the individual. Furthermore, this conception of the individual is intimately linked to capitalism. C. B. Macpherson who coined the term "possessive individualism," elaborates on the definition of property and its relation to individualism:

> Man [sic], the individual, is seen as absolute natural proprietor of his own capacities, owing nothing to society for them. . .Everyone is free, for everyone possesses at least his own capacities. . .Society is seen. . .as a lot of free individuals related to each other through their possessions, that is, related as owners of their own capacities and what they have produced and accumulated by the use of their capacities. . .Finally, political society is seen as a rational device for the protection of property, including

capacities; even life and liberty are considered as possessions, rather than as social rights with correlative duties.[18]

In the logic of liberalism one owes nothing to society for one's capacities, and, therefore, one can not appeal to society for aid in altering one's capacities (note, this is on the level of the individual and not the institutional). The individual thus stands as separate from society and has only him or herself to rely on.

Hence, the increasing circulation within the studios of commercial products such as self-help books, survival kits, audio-tapes, or videos. These have in common a strong belief in individuals' inner strength and capacities, both mental and physical, to resolve the crises in their lives. In the process, such an approach depoliticizes social ills by translating social crises into personal problems. In the end, the progressive elements that the shows might entertain become neutralized and eventually dissolve through the discourse of self-reliance for more compliance.[19]

The mother of a drug addict, an out-of-control teen, or the victim of a racist employer are given guidance to search within themselves for those "ingredients" that will momentarily ease the pain without disturbing the system of oppression that perpetrates the injustices. This is because virtually all programs find flaws in the individual, and these flaws are defined and contrived by standardized conventional "public knowledge." The expert brought to the stage of the show is there not as an emissary of democracy nor does s/he bring to the program a politically meaningful project of intervention. S/he is there as a professional advisor who will "empower victims" with survival strategies that engage no more than the individual present. The expert is in other words the voice of the status quo and the spokesperson of the doctrine of individualism: "if anything is wrong with our lives, it's because we have not tried hard enough." This logic takes several expressions and forms on the shows.

Research on talk show viewers provides empirical evidence demonstrating that "heavy talk show viewers were more likely to perceive that individuals have the power to solve these problems" where these problems are loosely defined as social issues represented by deviant behaviors.[20]

Another example is from the "Change Your Life" expert, Dr. Phil "Tell it like it is" McGraw. We first meet Dr. McGraw on January 13, 1999. Referred to as a "Life Strategist," McGraw comes on Oprah to discuss his ten life laws to get what you want and to promote his new book. Since that January, McGraw has been a regular on *The Oprah Winfrey Show*. In fact, he appeared every Tuesday in the "Get Real Challenge" in Oprah's 2001–2002 season and has now a very successful talk show of his own. The magic of McGraw, what has perhaps made him an Oprah favorite, is his credentialed common sense. While he doesn't have the "authenticity" of Vanzant, he does have a way of reconciling therapeutic and liberal discourses with audience common sense. In fact, compare the following quotes. The first is from Abt and Seesholtz who are against the talk show as a legitimate therapeutic site. They argue against the idea "that pop psychobabble or trite homilies or generalized panaceas (e.g. developing 'self-esteem') will repair the failures brought about by illness, economic deprivation, educational inadequacy and other cultural conditions."[21]

Similarly, McGraw says of himself, "I'm not much on psychobabble and buzzwords and letting your inner-puppy come out and pee or whatever. . .That's just not me and I'm not for everybody" ("After the Show" interview, www.oprah.com). He claims a position seemingly counter to the professional therapeutic discourse. However, his position in both the professional and the public realm is reinforced by Winfrey when she says of McGraw, "Common sense. That's what I think a lot of people are missing. That's why I like him, not because of the degrees, but the common sense"[22] So, it seems that the talk show has found its perfect expert—one grounded in common sense but credentialed through professional discourses. Dr. Phil is at one and the same time the host, expert, and embodiment of common sense. With his widely popular show, he has ushered in yet another evolution in the talk show programs.

So what is it that "Tell it like it is" McGraw says? In an episode on anger management (9–21–99), the focus turns towards two white men. Here anger is defined as rage and violence. Anger becomes confined to the male world. Instead of raising larger questions of patriarchy, power, and control, Dr. McGraw offers home-spun wisdom like "You gotta rise above your raisin'" (definitely not psychobabble). In these instances, domestic violence is not explored as a national epidemic, but featured as a manageable personality flaw ("Life is managed, not cured"). Specific and individual solutions are given to solve each man's problem. There is a general understanding that anger and violence are grounded in power and control. But, these are questions located in family structures, another social and discursive construction, and not in the social construction of masculinity. The guests are then given strategies to reroute their anger. Dr. McGraw even reminds them that they're the adult and they can choose whether to be angry and how to react to that anger. In the talk show tradition, the construction of patriarchy and the problem of domestic violence are social issues once again narrowed, focused, and fixated on an individual where they are easily solved without a whole lot of mess. "Telling it like it is" and "getting real" obviously mean once again reinforcing a liberal and therapeutic individualism.

The notion of talk shows as therapy is perhaps one of the more contested claims of the genre. Some members of the psychology community argue that "the shows fulfill a positive social function and are a legitimate source of psychological advice,"[23] while others would argue that talk show therapy undermines the value of actual psychology/therapy and trivializes the discipline as a whole. At the core of this struggle is the power to constitute a therapeutic discourse in a commonsense way. The results of a recent focus group research suggest that talk show experts are not really taken too seriously. One talk show viewer noted their function as legitimizing the show. Another sums up the role of the talk show expert as "She talks to people for 3 minutes, figures out their whole life, tells them what to do and then pats them off the stage. . .its just seems so fake." It is these moments when there is disharmony between the therapeutic discourse and the spectacle.

In this way, we can see how the talk show would reject the official therapeutic discourse in favor of a commonsense therapeutic discourse. This brings us back to the Carpignano et al. quote; "The talk show rejects the arrogance of a discourse that defines itself on the basis of its difference from common sense" (52). Masciarotte takes it a step further when she writes, "When the talk show rejects the institutional expert's conclusive voice (i.e. PhD's, congressmen, doctors), it privileges 'ordinary' experience, the 'authentic' voice of the everyday people, or street smarts of the working class" (86). In this way, talk show experts have to make sense to the audience for their professional therapeutic discourse to merge with the common sense discourse of the audience.

This was done on *The Oprah Winfrey Show's* "Change Your Life" series. What is perhaps most interesting on the series is that the guests/experts become the topic of the show—their therapeutic discourse becomes the show. The "topic" is merely the frame through which this discourse will be presented. Talk shows want us to believe that the presence of an expert on the show is not only desirable but increasingly unavoidable as well. The skills s/he will bring to us are beneficial to all, and the advice offered will bring happiness if only acted upon diligently.

Hence the experts' "license" to define people's mistakes and label their misdeed. In the process their power is given an opportunity to express itself. They acquire the license to do things others are clearly denied: establish the moral right and wrong of the guests, and act as guardians of the dominant norms and values in society. In brief, they are granted the power to define the very terms through which people should think about themselves and organize their lives.

This is how we can explain the existence of unequal relation of power on daytime talk shows. The programs' guests are categorized, through their very presence, as people with deficiencies and incompetences, while the professional experts are invited as help-providers who will "restore" order and "normalcy" to the social system. Experts are mostly expected to preserve the smooth functioning of the institutions in the social order (marriage, parenthood, schooling, etc.). From this perspective the "self-help ethos of talk shows may be teaching women more about how to monitor their behavior within the prescribed norms of American culture than about how to empower themselves." [24]

At issue also is not simply self-actualization but "self-regulation" since the shows are important spaces where one can begin to see the invisible hand of power at work, and ways in which various forms of knowledge provide the foundation for bureaucratic control. [25] This is how we can better understand the close links between vulture culture and the talk shows.

TYRANNY OF EXPERTISE OR THE LIMITS OF COMMON SENSE?

Researchers are becoming increasingly skeptical about the degree of interference of expertise in our lives. In his book, *The Tyranny of Experts* Jethro Lieberman makes a strong argument about how professional experts have become serious power wielders who increasingly narrow down people's choices and decisions in life. They have gradually but overwhelmingly taken hold of our daily existence by assuming that the citizen, like the consumer, is incapable of making meaningful decisions affecting his/her life. "We have turned over to others the power to make legal, medical, aesthetic, social even religious decisions for us. We have put the experts in charge— or at least abdicated our responsibility to them."[26]

Although Lieberman's descriptions of the scope of power of modern expertise are somewhat alarmist, they nonetheless urge us to rethink the heavy presence of expertise on the daily programs. They also alert us to the complex expressions of power relations on the multiple daytime talk shows and analyze the dimensions of power they embody. We are also encouraged to relate the shows' obsession with therapeutic advice and identity crisis to the wider context of social, cultural, and economic changes characterizing technologically advanced societies.

Without situating these programs within their larger environment, we would fail to understand the close link between the shows' obsessive preoccupation with the self and the "culture of narcissism"[27] we currently live in. Put differently, we need to understand the basis for such an exaggerated interest in individual lives, both by the shows and their viewers. Otherwise the programs would remain a mere spectacle of "Is" and "Ohs!" that is at best meaningless and at worst tedious in its daily repetition.

Concern with personal narratives and the obsessive need to publicly authenticate oneself and one's emotions illustrate some dimensions of the malaise characteristic of the conditions of living and being in late modern societies. Watching daytime talk shows is a confirmation that we are living in a "narrative moment" where there is a clear proliferation of personal narratives of the intimate.[28]

The telling of stories, however, is not an innocent practice, nor is it devoid of power relations and dimensions. Who tells the story, to which public, in which context, and through which medium are important levels at which stories and power are intertwined. There are hierarchies and patterns of power that determine the timing for the story telling and the kinds of stories that will be told or aired and those that should be silenced or ignored. Power punctuates virtually all the stages of the narrative moment. Power and stories are intertwined: power flows through them, makes them speak or silences them.[29]

On the other hand, the narrative moment signals an important social phenomenon. Individuals seem to be constantly in search for self-identity and increasingly finding it difficult to establish meaningful social relations. The "trials and tribulations of the self"[30] in search for ways to better connect with its environment is a feature of contemporary societies in much the same way as the numerous self-help products that promise to solve these tribulations. This should not surprise us given the fact that support networks for strengthening relationships, and developing or maintaining human intimacy are increasingly vanishing under the pressure of modern life styles and demands. Human bonding and moral support that used to provide individuals in precapitalist, premodern societies with the necessary advice in times of crises, have disintegrated. They have left a void that is being increasingly occupied by the often-paid expert advice and professional help. Thus, "the break up in the protective framework of tradition" has made therapy "a secular version of the confessional."[31]

The "culture of narcissism" to borrow Christopher Lasch's term, points to the pervasiveness of the therapeutic sensibility in America due to the increasing anguish and isolation of the people. Narcissism in his analysis means a high degree of dependence of individuals on the technical skills of the certified experts. The dependence of some is such that they cannot organize their family life, raise their children, or plan their career without the help of certified expertise/experts. The greater the dependence, the lower is their self-esteem and insecurity.[32] What is daily offered on the talk shows is but a visual expression of the angst and agonies of the modern self. The programs stage an everyday display of the high levels of "reflexivity" that characterize our times and the therapeutic mood that accompanies it.

Reflexivity is not harmful in and of itself: it is critical for personal growth and maturity. However, put within the context of therapeutic excesses, it does not liberate the individual from the obsession with the self. On the contrary, it encourages him/her to focus more and reflect longer on the isolated self, stripped from its larger social context. At the same time, such an obsession opens rooms for expertise to step in private lives and take hold of it in a legitimate fashion. Self-reflexivity of late modernity, as Giddens puts it, "extends into the core of the self... the self becomes a *reflexive project*."[33]

Daytime talk shows are clearly positioned within the context of reflexivity as detailed above. These are media spaces for producing, sharing and circulating knowledges be they personal, intimate, or otherwise. Daily shows are the popular sites where the centrality of personal narratives is as important for the survival of the programs as the circulation of therapeutic knowledge. Although the shows' experts claim a scientific basis for their approach to the guests' problems, and despite their endeavor to "empower" the oppressed and the marginalized, their approach does not go beyond making them cope with their own sense of victimization.

In other words, and as this section revealed, there is an unequal relation of power between the shows' guests and experts that the programs work towards reinforcing and maintaining. For the sharing of personal experiences has nothing "democratic" about it if it is addressed exclusively in psychological terms. Sharing invariably requires a political and ideological analysis, and personal dramas must be understood within a social praxis that entails both meaningful reflection as well as action. Talk shows might give lay people the opportunity to voice their concerns for a few minutes each day, but these programs are in no way a platform for developing a political project that aims to dismantle oppressive structures and mechanisms. Acting out fantasies or frustrations on the stages of the shows, be they that of *Jerry Springer*, *Maury*, or *Leeza*, might be seen as no more than a dramatization of what social conventions judge as "the permissible limits of anti-social behavior."[34]

What should also be addressed at this stage is the role of the show host, his or her participation in the struggle over knowledge, interpretation, and power on the shows.

SHOW HOSTS: STARDOM AS EXPERTISE

The role of the talk show host is crucial. As a moderator of voices and discourses, and as the facilitator of the production of knowledge and its circulation, s/he occupies a central position in the daily programs. The host constantly manipulates life stories and personal narratives, and processes the common sense and expertise of all his/her guests—professional experts, participants, and studio audience. In doing this, s/he not only reveals a remarkable power to "validate" the narratives and positions of others, but s/he mainly establishes an aura of authority for him/herself.

But, what makes talk show hosts enjoy so much power and freedom to discuss people's personal lives with so much confidence and seeming competence? What knowledge bases do they draw from, and what sources allow them to gain the status of "expertise"? Does Oprah Winfrey have more authority on race issues, for instance, than Ricki Lake because of her cultural capital as an African-American woman? Does Jerry Springer have more authority to discuss Nazism and fundamentalist groups because of his capital of oppression as a Jew? Do we expect O'Donnell and Winfrey to empathize more on weight issues than Sally Jesse?

Advertising statements for Vanzant's own show in 2002, for instance, seem to suggest that the host's cultural capital is an unquestionably strong basis for establishing the authority of the hosts in specific areas of knowledge. Vanzant is quoted as saying: "The challenge for me was making that shift—understanding I didn't have to wear my culture, but I had to embody it," she says. "I didn't have to talk about it, I had to demonstrate it."[35] The queen of the daytime shows, Oprah Winfrey, was a pioneer in establishing the tradition of using the capital of oppression from her personal/private life to further her professional success. Helping Vanzant in her debut, Oprah makes the following statement "And if a wretchedly traumatic past is a requirement for being a good talk show host … Vanzant is way overqualified."[36]

We learn then that Vanzant was abused by relatives and spouses early in her life, Vanzant (born Rhonda Harris) was once a "walking catalog" of personal crises and dilemmas: teenage mom, welfare queen, drug abuser, and suicide survivor.[37] Yet, by the time she decided to become a show host, Vanzant had already amassed an

impressive community of readers and fans. Three of her 10 books—
"Yesterday I Cried," "In the Meantime" and "One Day My Soul Just
Opened Up"—have hit the *New York Times* best-seller list. Success
seems to breed success: her success on the *Oprah Winfrey Show* has
played a major role in the ease with which she is gathering readers
and fans. While her show failed to keep up with the competition
(possibly due to format problems) and was cancelled, she has re-
emerged in a context more suited to her talents—as the life coach on
the daytime reality show *Starting Over*.

The life of Sally Jesse Raphael reveals an equally unique cycle
of tragedies and triumphs. The first include living on food stamps,
experiencing the brutality of rape (the victim being her mother),
the tragic loss of her oldest daughter and near death of her son. Her
successes include national recognition for her exceptional skills as a
listener and advice provider on her syndicated talk radio show.

What these few examples of the show hosts' biographies reveal is
that their personal tragedies and traumas are not different from the
personal stories and private experiences they discuss on their shows.
We are repeatedly reminded by most show hosts that they are like
us, and their lives are similar to ours. A human bond is established
that declares us all fellow sufferers, hence, we can relax and tell it
all on the stage of the show. In particular, we are invited to trust the
show hosts to guide us in life and empower us with survival strategies to move on.

But the similarities between the host's life experiences and the
guests' personal tragedies often end in the relatively common experiences of pain. The present and the future of the host and the guests
are very dissimilar. Talk show hosts are stars in the media culture.
Most hosts were either determined to rise to the status of a star or
flirted with stardom prior to becoming a regular fixture of the daily
shows.

In most cases, they brought with them their stardom to the
shows and built further prestige on it. Their comfortable status of
stardom gives them, for instance, an additional dimension of power
and legitimacy to evaluate the guests, judge their beliefs, and sanction their attitudes. In a word, show hosts are modern experts of
another breed.

Leeza Gibbons, for example, "brings her personal mix of style, compassion and humor" to each show, at the same time she "draws on her own experiences as a working woman, wife and mother." Montel Williams, we are told in his biography, comes to the competing arena of daytime show "with an incredibly untraditional background."[38] He is an honorably decorated former naval intelligence officer who established his reputation as a renowned motivational speaker prior to starting his daily show. At the height of his success, the discovery of his multiple sclerosis has given him additional experience to build further expertise around. Ricki Lake, on the other hand, is the "versatile actress and performer" who was already in the public eye for a while. She starred in numerous films including *Mrs. Winterbourne*, *Hairspray*, *Cry Baby*, and *Serial Mom*.

These additional examples from the lives of daytime show hosts reveal the capital of both pain and fame they bring with them to the programs. Although some show hosts are more willing than others to share childhood traumas and tragedies with the public, most of them show little hesitation in weaving their personal experiences into the professional persona. Their confidence to talk about virtually all subjects with the voice of a competent expert comes precisely from this status of stardom. Stardom, itself, is obviously media-constructed and fabricated according to the logic of the market.

Very much like the stars of the film industry, show hosts represent a vital element in the economics of the talk show world. Their life experiences prior to hosting shows constitute a "capital" as well as an "investment" for the success of the shows: the more dramatic and "unconventional" their lives are, the higher the chances of their survival in the ruthless world of daytime shows.

It is not only important to underline the economic dimension of the personal/professional aspects of the hosts, but also the privileged position they occupy in the defining social roles as well as patterns of behavior. Richard Dyer, whose work on the phenomenon of stardom is insightful, reminds us that stars rarely expose or embody attitudes that are contrary to the dominant ideological values in society. [39] Stars are the guardians of the status quo because it is this status quo which allows them to progress and prosper. Perhaps the most telling example of the hosts' reinforcement of conservative values is the popularity of such disciplinary measures as "boot camp" among the show hosts.

Armed with an established fame then, and sustained by a media-constructed popularity, the hosts have little difficulty in establishing themselves at the level of "expertise" and as such, their knowledge often competes with if not trumps that of the participants or the guest-professional for greater legitimacy. The hosts often invite others' views or encourage different contributions only to bring closure to a discussion by re-establishing an order of perception or understanding in which his/her words stand for "the" last word.

The following example provides an interesting combination of expert knowledge and commonsense knowledge through the persona of show host. We are introduced to Iyanla Vanzant as an author and businesswoman. In an Oprah episode airing January 4, 1999, we meet Vanzant and learn of her credentials not only as an expert in the professional sense, but also as an expert of common sense. We learn of her life story as a survivor of an abusive childhood and marriage and her struggles as a black woman trying to get out of the welfare system and get an education. Beginning with those credentials, Vanzant comes to speak from a position of unquestioned authenticity—one that can only be claimed by straddling the line between common sense/practical knowledge and professional knowledge.

Using quotables such as "Movement begins with changing your mind" and "The only thing God asks of his children is to make the best of what you got," Vanzant teaches how to live and succeed in the prevailing power structures. Under this strategy, one must reconcile oneself to accept the terms of "the world" rather than challenge those terms. Changing lifestyle and circumstance becomes as simple as changing your mind. And, the "make the best of what you got" comment echoes the liberal individualists' view that our own abilities are all we need to succeed, regardless of race, class, or gender.

In the same program Vanzant attempts to empower her audience by saying "you are not what happened to you." In this way, she hopes to allow the members of the audience to redefine themselves. But, I am left wondering, if I am not what has happened to me, then what am I? Of course, the answer is that I am a free and autonomous individual who is capable of making decisions outside of, or rather despite, the institutional and discursive agencies that frame and structure my life. In this way, Vanzant removes the individual from the social, political, and economic context that created her. This is not to say that Vanzant does not acknowledge and understand racial and economic inequality. However, she chooses to chalk up personal experiences of injustice to a spiritual journey by saying "Wherever you are in life is where you're supposed to be." And, she makes a convincing argument. After all, she offers herself as an example. By following her own philosophy (and thanks to her appearances on *Oprah*), she has become a best-selling author, hosted her own short-lived talk show and is now featured as a life-coach on *Starting Over*. However, this only reinforces the individualistic nature of success in a capitalist-democratic society. It also obscures the true power influence of social and economic institutions. Without a doubt, in the United States, commonsense logic naturalizes the idea that individuals have the right to self-determination and freedom of choice. We all make choices everyday which affect our lives. But, we must concede that our lived experiences are framed by socioeconomic circumstances. We must remember to put everything in context so that we recognize that the color of our skin, our gender, our social class, and our sexuality all impact our view of the world *and* the way the world views us.

With so many claims to expertise, knowledge, and truth, American daytime talk shows give a regular display of "the hierarchy of credibility."[40] They have altered our perception about who has priority and legitimacy over credibility. Without a doubt, stardom in American society is granted a privileged position particularly when it reduced the logic of "expertise" in the language of "common sense." Talk show hosts are perhaps the most popular experts media culture has produced.

In conclusion, talk shows capitalize on the "scientific" knowledge of experts, the common sense of the programs' audience/guests and to a large degree on the stardom/expertise of the programs' hosts. This chapter presents talk shows as concrete cases demonstrating the intersection of common sense and expertise through which people make sense of their everyday lives. Vulture culture is critical in shaping this intersection: it orchestrates and charts the ways individuals should use these knowledges and have them guide their lives. Ultimately, vulture culture determines which knowledge is valid and valued. One of the ways this is done is through a spectacular staging of confrontation between various claims to knowledge. The next chapter focuses specifically on the spectacle aspect of vulture culture and talk shows.

5

ELECTRONIC CARNIVAL:
SPECTACULARIZING TALK

Daytime television talk shows have invited a mixture of resentment and appreciation from the general public ever since Phil Donahue first introduced the format in 1968. Celebratory comments about the democratic spirit of talk shows consider them the "last bastion of freedom of speech" where the voices of the ordinary, powerless, and underrepresented citizens can still be heard.[1] Condemnatory criticisms of the shows, on the other hand, view their irredeemably exploitative and sensational nature as a threat to the very health of democracy.

Yet, whether audiences entertain a love or hate relationship with talk shows, the programs have grown to be commonly referred to by critics and supporters alike as the "geek" or "freak shows," the "sicko circuses" or the "carnivalesque space."[2] The labels are often used interchangeably to capture the essence and the spirit of what talk shows are considered to be about. They are the world of the trash and the tasteless, the vulgar and the vicious, the grotesque and the offensive, the silly as well as the spectacular. In a word, talk shows are increasingly seen to be no more no less than programs where the "freak" and the "geek" fret and strut on the stage of a televised show, which they exit after brief moments of sound and fury—signifying nothing.[3]

In fact, even the "King of Trash TV," as Jerry Springer is now called by the media, finds it hard to contest the labels attached to

him and to his show. This is what he has to say to the viewers of his
58 minutes *Jerry Springer: Too Hot for TV* video: "You know, I'll be
the first to admit that we have got a pretty crazy show on our hand
here. Sometimes I can't believe it myself ... remember it's a crazy
world ... sometimes our guests just go too far ...some are too outrageous." For Steve Wilkos, Jerry Springer's chief security on the
show, Jerry is no outsider in this circus: "he is a little geek too," he
admits in a televised biography profile of the show host. Is this the
whole tale, though?[4]

We argue in this chapter that the world of talk shows is much
more complex than a simple "tale told by an idiot," and much more
ambiguous than a mere entertainment spectacle. By focusing precisely on such key concepts as the "carnivalesque" and "the spectacle," we expose some of the key aspects of vulture culture. We reveal
how talk shows are, in both form and content, ideologically messy
and ambiguous enough to offend as well as appeal to a wide range of
viewers from very dissimilar backgrounds or allegiances. Talk shows
will be discussed, then, from the perspective of a constructed space
for the spectacularization of talk, the performance of private experiences and the staging of social conflict.

The first section of the chapter discusses the Bakhtinian notion
of the "carnivalesque" and the interesting ways in which it can be applied to the study of popular entertainment genres such as daytime
talk shows. This brief theoretical overview also provides the necessary background to explicate the ritualized theatrics involved in the
staging of both action and talk on the shows. Aspects of role-playing
and the blurring of boundaries between media genres in the shows
format are analyzed in order to reveal the multiple forms of exploitation and oppression inherent in the very structure of a daily spectacle. The final section of the chapter discusses the carnivalesque
appeal of talk shows from two specific angles: the spectacle of the
out-of-control youth, and the spectacle of women-as-victim/wom-
en-as-transgressor. The selection of these two specific instances is
meant to provide a closer look at how a "mere" spectacle betrays
serious sociocultural definitions of age, class, gender, sex, and ethnic
differences.

Daytime talk shows have co-opted the entertaining aspects of
the early popular carnival, and readjusted them to the contemporary

logic of media ratings as well as market profits. The shows' capitalization on the spectacular dimension of the carnival neutralizes the potentially transgressive or liberating forces these spaces of communication could provide. The critical perspective we adopt in this chapter will, thus, reveal that despite the shows' frequent appeals to the tenets of democracy—justice, equality, and freedom of speech—their considerable success depends largely on their reinforcement of cultural, economic, and political oppressions. A critical pedagogy of talk shows, as we define it throughout this book, helps to unravel precisely those aspects of the daily programs which might be spectacularly entertaining yet insidiously oppressive.

CARNIVAL IN THEORY

A discussion of the carnival or the carnivalesque dimensions of popular culture invariably invites an overview of Mikhail Bakhtin's inspiring work on the subject.[5] Bakhtin's *Rabelais and His World*, does not simply define the concept of "carnival" during the French Renaissance, but also explores the complex relationship between popular forms of entertainment and the established social order. In this respect, his insightful conclusions on the subject provide a theoretical model flexible enough to allow the analysis of newer forms of carnivalesque experiences in our electronically mass-mediated age.

Bakhtin organizes the manifestations of classical folk culture in at least three distinct forms: a) *ritual spectacles*, which include pageants, comic shows of the market-place, popular fairs, feasts, processions, competitions, dancing, amusement with costumes, masks, giants, dwarfs, and monsters; b) *comic verbal compositions*, which include parodies in both their written and oral forms; and, c) what he calls various *genres of billingsgate*, ranging from curses, oaths, and profanities to popular tricks as well as humor in their crudest form.

Bakhtin pays particular attention to two aspects of the carnival: the "grotesque body" and the "carnival laughter." Briefly defined, the grotesque body, which is closely linked to folk humor, refers to a whole set of images of a body which "transgresses its own limits" through its exuberance, excessiveness, and exaggeration.[6] Carnival laughter, on the other hand, is as he explains:

[N]ot an individual reaction to some isolated "comic" event. Carnival laughter is the laughter of all the people. Second, it is universal in scope; it is directed at all and everyone, including the carnival's participants... Third, this laughter is ambivalent: it is gay, triumphant and at the same time mocking and deriding. It asserts and denies, it buries and revives.[7]

Further, Bakhtin insists that the carnival is not merely an entertaining spectacle to be watched by detached spectators, but a culturally bustling space in which all people are expected to participate regardless of the socioeconomic, cultural, or political differences they may have in real life. In other words, the carnival becomes the topsy-turvy world where established hierarchies are temporarily dissolved, conventional norms and codes of behavior are suspended, and human deepest passions as well as tensions are released. The logic of the carnival celebrates (low) popular expressions and experiences, and it pays no tribute to the knowledges, values, or expectations of (high) official culture.

In this sense, the carnival's temporary suspension of all hierarchies, ranks, and privileges throws into relief precisely those dividing lines existing between the powerful and the powerless, the privileged and the marginalized in society. The study of the carnival enables, thus, a better understanding of the power relations between different social classes, cultures, and languages existing within any social order.

Bakhtin's celebration of the early forms of the carnivalesque space, however, easily slips into an unqualified enthusiasm about their transgressive or liberating potential. As many critics have already observed, the "relief" from repression and social control which the popular carnival seems to provide, is in fact temporarily "licensed" by official powers in order to be better "policed" by them. Roger Sales makes the point succinctly when he states:

There were two reasons why the fizzy, dizzy carnival spirit did not necessarily undermine authority. First of all, it was licensed or sanctioned by the authorities themselves. They removed the stopper to stop the bottle being smashed altogether. The release of emotions and grievances made them easier to police in the long run.... The Carnival could be a vehicle for social protest and the method for disciplining that protest.[8]

This dual aspect of "licensing" and "policing" is part of a larger debate over whether carnivalesque spaces, in their classical or contemporary forms, are politically conservative or progressive.[9] Important as these considerations are, they do not however refute the fact that carnival is infused with political meaning and tension, nor do they undermine its validity as a conceptual tool for understanding conflicts and contradictions in the cultural and sociopolitical landscape of modern societies. As Peter Stallybrass and Allon White cogently argue in *The Politics and Poetics of Transgression*, carnival is "intrinsic to the dialectics of social stratification"; and as such, it remains useful in its application to different cultural manifestations in different historical periods.[10]

Drawing on Bakhtin's early work, and extending his interest in the relationship between the "high" and the "low," these researchers demonstrate that these categories—whether applied to classes, cultures, knowledges or lifestyles—have been persistently used throughout European civilization to distinguish between the official culture of the elite and the popular culture of the marginalized "others." But the relationship between these two terms is more problematic than a systematic categorization might suggest. That is, although the "low" is loathed and feared by the social and political elite, it nevertheless constitutes an important social element upon which the upper class depends for maintaining its sense of distinctiveness, and validating its power as well as privileges.[11] The ambivalent attitude which the official culture has shown towards low classes results in, what the researchers call, a "mobile, conflictual fusion of power, fear and desire in the construction of subjectivity: a psychological dependence upon precisely those "Others" which are being rigorously opposed and excluded at the social level. It is for this reason that what is *socially* peripheral is so frequently *symbolically* central."[12]

Our critical approach to daytime talk shows, in this chapter as in the rest of the book, proposes to see these programs as visual expressions of vulture culture. They occupy central spaces despite the peripheral position American society would prefer to give them. First, there is little doubt that the general reception of daytime talk shows places them at the lowest level of the debasing and "vulgar" forms of modern popular entertainment. Labels such as "sicko circuses" or "freak shows" contain a strong normative assumption about their "low" standing in the world of entertainment. They do tell us little, however, about the political implications behind the construction or use of these labels. Again, it is the carnivalesque and populist nature of talk shows which makes them a source of both attraction and repugnance to their ever growing diverse audiences. But, to simply declare daytime shows "vulgar" without questioning how this vulgarity is constructed and orchestrated by the dominant power structures in society, is to simply add further confusion to a heavily charged term. Important clarifications, therefore, need to be made at this stage before we explore ways in which daytime talk shows recreate the carnival world under an electronically modern guise. First, it is important to emphasize that vulgarity is not a *given* category in the social world, and much less so in the media culture. Vulgarity is, rather, a *construct* manufactured, manipulated, and exploited by the dominant institutions it benefits the most.[13] In the case of talk shows specifically, vulgarity cannot be discussed outside the general context of power relations and exploitation within which the programs are produced and consumed.

We have sufficiently demonstrated in chapters 2 and 3 the economic imperatives behind the special infatuation of the major television networks with daytime talk shows. Without belaboring the point, let us just restate the fact that the daily shows are among the cheapest programs to produce yet the most lucrative source of profit for the media giants. What needs to be underlined here is that the real source of profit is, in fact, nothing else but real-life stories and human crises or choices turned into a series of daily spectacles of "freakiness" or "vulgarity." That is, if the spectacle is to entertain, humor and attract its audience's attention, it has to appeal to those aspects of social and cultural life, which are already labeled by the dominant social order as "low" and "vulgar."

In this context, differences in gender, class, race, age, ethnicity, lifestyles and sexual orientation become the most valued items for constructing the electronic parade of carnivalesque "oddities." And differences are valued not because the programs are politically committed to dispelling the prejudices or misconceptions surrounding them in the established social order. Rather, differences are valued because they make good material for a spectacle: they can be easily staged, spectacularized, mocked at or ridiculed.

Having said this, the manufacture of "vulgarity" on talk shows is not a pure exercise of force or coercion from which the shows' guests and viewers emerge as hapless victims. No talk show is or can be produced without the consent as well as the active or vicarious participation of its audiences. And this is an important point that will be illustrated in a later section of this chapter. Suffice it to say at this stage that we consider the shows' participants and viewers to be neither "dupes" nor "naïve" victims of a large-scale conspiracy mounted against them. Their relationship with the different shows partakes of a complex dialectic in which both consent and exploitation feed into each other. That is, the programs' viewers and participants may choose the degree of their involvement in the shows, but such a choice or participation does not exclude unequal relations of power between the shows' producers, its participants, and viewers.

Thus, far from being inconsequential or "vulgar" entertaining episodes, we need to see talk shows as a confirmation of existing social conflicts and antagonistic sociopolitical forces. Above all, we need to see them as manifestation of a larger culture we call here vulture culture. Daytime talk shows constitute a carnivalesque space because they are situated at the intersection of the high and the low elements of cultural life as well as social experiences. And as such, they represent the spectacular stage upon which the confrontation between the "central" and the "marginal" elements in society take place, and the dissonances as well as disjunctions in the social fabric are spoken to. This is done in the very process that these programs spectacularize, trivialize, and capitalize on human drama. Fear, desire, and power are indeed related to these communication spaces which are often deemed devoid of political intensity. The following section focuses on how these tensions and tendencies are played out on the stage of the shows.

INSIDE THE TALK SHOW CARNIVAL

In an advanced capitalist society, where most human experiences and patterns of communication are increasingly mass mediated, the carnival takes on an electronically mediated form as well. Daytime talk shows have grown to constitute, through their diversity of contents, concerns, and formulas, the cultural space where the carnival experience is both revived and reviled. Talk shows appeal to our collective memory of popular festivities and resurrect images of primal human passions associated with them. They do so because they are largely built on and further stimulate the expression of such basic human sentiments as desire, jealousy, pity, hatred, anger, shame, meanness, and cruelty. The spectacle that is constructed daily seeks to engage its viewers through a successive parade of emotionally charged images, experiences, and narratives drawn from the complex social and psychological landscapes of our modern lives.

More than any other televised programs, perhaps, talk shows have revealed since their inception an incredibly voracious capacity to borrow from and incorporate a wide range of entertainment genres with which we are familiar. In both form and content, the programs borrow, for instance, from the tradition of oral story telling, theatrical performances, popular melodrama and soap operas. In this process of continuous borrowing and fusing of genres, they demonstrate vulture culture's ability to constantly reinvent and reposition itself. They also try to recreate the democratic atmosphere of the town hall gatherings, yet they appeal to the entertaining qualities of the circus exhibit of the "grotesque" or the "odd." And, lest we forget the latest re-incarnation of talk shows with *Jerry Springer*, there is an increasingly strong presence of the rowdy and mighty world of professional wrestling. It is this eclectic nature and elastic formula of talk shows which holds the promise of recreating a carnival space with a transgressive impetus. Yet, the same defining aspects of the shows make them elusive in their ambition and contradictory in the knowledges they produce. Specific examples of these contradictions will become clearer as we briefly comment on the different genres incorporated in talk shows.

First, talk shows contain residual remnants from the earlier carnivalesque world, as discussed earlier by Bakhtin, because they constitute that "uninhibited" space for releasing tensions and discharging accumulated or buried frustrations. These periodic moments of emotional release take place within a repetitive time frame which reminds us, albeit in a distorted fashion, of the carnival's cyclic perception of time. The seasonal repetition of popular carnivals was the basis on which participants re-established and re-kindled their sense of belonging to the larger community and social life.[14] In the case of talk shows, what was once a seasonal event has been turned into a daily routine. That is, our electronic carnival has been readjusted in format and intent to meet the market logic and commercial imperatives of the media and corporate culture. Talk shows' repetition, in this sense, is measured by daily ratings; while the viewers' and participants' involvement is translated into accumulated capital. In a word, talk shows' time is money.

Second, talk shows recreate the transgressive spirit of the carnivalesque world because they do not shy away from experiences or narratives usually deemed too personal, shocking, offensive or tasteless by society's official canons of acceptability. On the contrary, the programs constantly probe into the deepest recesses of human lives and emotions to "show" that which is normally hidden and "talk" about that which is somewhat forbidden. Common topics repeatedly approached by talk shows include, for instance, unfaithfulness, kinship feuds, domestic violence, rape victims, obsessive or compulsive behavior, rebellion of teens and fantasies of adults, questions of authority and control, sexual preferences and orientation. These issues are raised within a communication environment that keeps the tradition of oral story telling alive, and the sense of communal concern present with it. This is so because the "talk" on the shows is largely based on the narrativization of personal stories and the encouragement of the community's immediate responses to them.

Third, and closer to the point made above, talk shows seek to recreate a dynamic space of interactivity and dialogue in which all are invited to actively participate: the programs' guests, audiences and hosts. They do so, for instance, by carefully recreating a friendly, if not hospitable seating arrangement which reduces the physical distance between and among the studio guests, moderators, and audiences. This blurring of the spatial boundaries is further enhanced by the increasing tendency of seeing the show host amidst the studio audiences with a microphone in hand soliciting their questions or responses. The roaming and probing eye of the stage camera, on the other hand, tries to reassure us that "in a sort of democracy of lighting everybody is brought on stage and given their share of illumination."[15]

Herein, then, lies the talk shows' promise to re-create the democratic spirit of the town hall gathering where all the issues raised are worth debating and all the voices expressed are worth listening to. In other words, the format and the focus of the programs easily lend themselves to the creation of a progressive space where the experiential knowledge of the oppressed and the marginalized can be recognized as well as validated. The programs' emphasis on the interpersonal dimension of communication also makes them a potentially favorable environment for reflecting on the implications of personal issues in a critical, constructive, and collective manner. It should come as no surprise, then, that many talk show hosts advertise their shows through the very rhetoric of democracy and freedom of speech. Montel Williams, for instance, invites us to participate in his show with the idea of recovering "what talk *was* meant to be." Geraldo Riviera appeals to the concept of citizenship and civic duty through his "Speak up America!" reminders.

However, and having said this, talk shows have not built their reputation merely on the "telling" of human experiences or the validation of emotions, but also, and perhaps largely, on the re-enactment of these emotions in a theatrical fashion. The dramatic performance of daily narratives and the spectacularization of human emotions are unquestionably among the important dimensions on which the shows capitalize for their popularity. Thus, just as talk shows are a truncated version of the earlier carnivalesque space, their appeal to democracy has to be also seen within the overall media context where the stiff competition between the shows' ratings and their commercial profits can decide on a show's life or death. And part of the survival strategies imposed by the merciless logic of the media market is the exploitation of the sex-violence formula.

This is how talk shows bring together, in a rather spectacular fashion, both the democratic spirit of the town gathering and the action-packed world of professional wrestling. For, how else are we to account for the increasing popularity of daytime talk shows among different segments of society, despite the scathing criticism directed at them? And how else can we understand the trials and tribulations of the Jerry Springer show, for instance, which went from a spectacular failure in 1994 to a spectacular "success" in the last years of the same decade?

The answer to this question may be partly found in both the talk show in question and the career trajectory of Jerry Springer himself: from being an ambitious councilman and Cincinnati mayor, citizen

Springer has become the no less popular ringmaster of the wrestling-like talk show. The previous politician and anchorman, master of both the rhetoric of democracy and the world of performance, is again most likely to provide a plausible justification for combining the democratic spirit of the shows with the world of professional wrestling. In a passage worth quoting at length, Springer makes the following points:

> You know, we pride ourselves in showing you from time to time the most outrageous people of our society, those who are either wildly eccentric or those who in their political and social being are simply defiant of convention ... Now while none of these particular manners or life styles are the ones we would necessarily choose for ourselves, how boring life would be if there was none of this outrageousness, that is to say, none among us who would push the edges of the envelope. Now please understand, because we show it, it does not constitute an endorsement of it or any particular view or behavior, any more than reporting a murder on the news or prime time movie about a rape is an endorsement of those horrors.
>
> Look, television does not and must not create values, it's merely a picture of all that's out there: the good, the bad, the ugly. A world upon which we then apply our own values learned and nurtured through family, church and experience. Remember, if we permit only those views which the majority of us hold, then you and I are free only so long as we agree with the majority. If you believe nothing else I ever say in these commentaries I offer at the end of every show, believe this: The politicians or companies that seek to control what each of us may watch are far greater a danger to American democracy and our treasured freedom of speech than any of our guests could or will be.[16]

One has to admit that the above statements underline many of the serious contradictions embodied in the talk show formula and reflected in the public's responses to them. But, the appeal to view talk shows as "merely a picture of what is out there" is quite uncanny and certainly misleading. For if we accept such an argument, then talk shows are as ideologically neutral as pro-wrestling spectacles are. But we know that the world of wrestling is packed not only with action and drama, but also with the re-enactment of social intolerance to the marginalized and the oppressed.[17] Thus, it is only when we begin to explore issues of theatrical performance and rehearsal that we fully understand the ambivalent nature as well as ideological confusion of daytime talk shows. In other words, when we critically analyze the manufacture of "immediacy" and "spontaneity" of the talk on the shows, we realize that the programs have less enduring ties with the town gathering than with such entertainment genres as the soap opera and the popular melodrama.[18]

Let us pursue the parallels between these genres.

SOAP OPERAS AND TALK SHOWS

Talk shows remind us of the televised soap opera because both genres make family sagas and personal dramas the focal points around which action and reaction revolve. Both genres prioritize, in particular, family relationships with the complex web of secrets, loyalties, betrayals, as well as mysteries about and around paternity. Montel Williams, for instance, focuses his show in the first week of May, 1999 on "Accepting the Kid as One's Own" and "Mother/ Daughters: Secrets and Lies." The same week, Sally Jesse Raphael invites her guests to reveal how "Shocking Secrets Tear My Family Apart," while Jerry Springer frequently stages variants around "Paternity Tests Revealed" and "Holiday Hell with my Feuding Family." On her November show of the same year, Ricki Lake turns again to the issue of paternity secrets with "DNA test will prove you're my dad." The selection of topics on talk shows is clearly not limited to this list, but the melodramatic quality of family problems certainly accounts for their higher visibility.

What is also worth noting, here, is that these issues are raised on the shows, as they often are in soap operas, within a context that mostly polarizes good versus evil, hero versus villain, and moral versus immoral. Yet, while polarization enables the staging of confrontation in its various theatrical form, it gives us only a schematic, simplistic and trivialized version of human nature as well as the social world. Most importantly, it empties them of their depth and complexity, and then offers them as an entertaining spectacle for immediate production, instant consumption.

Having said this, and unlike soap operas' world of the rich and the famous, the problems which the guests and participants narrate daily on the talk shows come mostly from the lower classes and the underrepresented social groupings. They provide in this sense a more realistic rendering of social life with which audiences can easily identify or directly relate to. The shows' format and stage arrangements, as previously discussed, encourage audiences' involvement and immediate responses to the unfolding dramas of the day.

THE MELODRAMA OF TALK SHOWS

Thus, while the raging battle between good and evil is orchestrated on the stage of talk shows, the audiences' responses take form of an ongoing chorus of cheers, claps, laughter, gasps of disbelief or empathy, statements of approval or condemnation. It is in this sense that daytime talk shows remain more faithful, than soap operas, to the original tradition of popular melodrama. At the risk of oversimplifying the complex structure of popular melodrama, let us summarize here just those aspects of the genre which are useful to understanding issues of performance on talk shows.[19]

Some of the structural features of melodrama include: a) the polarization of the world and human nature into good versus evil; b) staging a central conflict between a villain and hero/heroine, who is rescued by the "figure" of justice; c) investing ethical/moral values in the physical appearance of the characters; d) the creation of spectacular "heightened effects" to carry on the narrative; e) the complicity of the audience whose emotions are appealed to on the basis of fear, enthusiasm, pity and laughter; and, f) the final resolution of conflict according to a "conscientious moralism" and respect for established norms.

The popularity and appeal of television talk shows may be explained in many ways by their reviving different melodramatic elements we briefly outlined above. In virtually all daytime shows aired daily, good and evil constitute the raging battle to be fought on the stage. The strong codification of the actors' appearance in popular melodrama are also present on talk shows in the sense that the moral character of the shows guests is often translated into the role they are expected to play for the day. Talk shows, it should be underlined, hardly ever promote their daily programs through the names of their guests. They call attention to the moral and ethical case to be discussed, and the guests become the embodiment of the case itself. Thus, the names of the show guests, as Charles Acland rightly points out, have a "metonymic" function: they are far less meaningful in and of themselves than the specific case they are going to act out.[20]

And "acting" on the stage of the shows is performed by virtually everybody: the host, the guests, and the studio audience. First, guests and audiences go to the show with a set of expectations about how and when they are supposed to act or react to the dramas in front of them. Behavioral expectations are acquired, for instance, through exposure to the programs prior to participating in them. Audiences are further rehearsed and prepared for action by the shows' technical crew. Also important to underline is the fact that guests and audiences associate specific behaviors with different shows. They adapt, as it were, their expectations to fit the Ricki Lake style of performance, or that of Rosie O'Donnell as they do with others.

It is in this sense that the audience can be seen as an accomplice in the unfolding staged dramas. Complicity is essential not only to the shows' success but to the audiences' sense of self-reassurance which is expressed in two ways. First, getting involved in the guests' story allows the audience members to either identify with the guests' feelings; hence, establish a sense of "community ties"; or, distance themselves from the guests; hence, confirm their difference to what is perceived as vilifying. Second, the audiences' complicity is expressed through their acceptance of the shows as the space for the gushy outpouring of repressed emotions and vocal expression of moral views. Show hosts often contend that it is the studio audience which becomes the "moral barometer" of what is proper and what is not on the shows. There is some truth in this statement. But the metaphor of the barometer is valid only if we remember that it functions with the invisible hand of the hosts and stage managers!

This is to say that show hosts also have a basic set of roles which they routinely rehearse and which audiences, in or outside the studio, have grown to anticipate. Some of these include, for instance, heightening the dramatic effect of the stories narrated, and coaching, if not monitoring, the reactions of the audiences accordingly. During the taping of one of *Ricki Lake* the audience was coached to applaud the "good" and "boo" the "bad." Without telling the audience what is good and what is bad, the expectation was that common sense and shared social values would be the guide. We also need to remember that most show hosts have a considerable familiarity with the world of performance and professional acting. That is, the stage with its technical and artistic devices has no secrets for them since most of them swing back and forth between movie acting, comedy performance, and show hosting. This is certainly the case of Ricki Lake, Montel Williams, Rosie O'Donnell, Oprah Winfrey, and Jerry Springer, for instance.

On the stage of the shows, however, hosts become simultaneously performers, moderators, experts, editors, and the figures of justice as defined earlier in popular melodrama. Show hosts play many roles, some of which are theatrical in form, while others are mostly normative in orientation. They are editors in the sense that when all is said and done, it is the host who decides on who will talk, for how long and during which segment of the show. The editing process is done mainly through the use of the microphone with which the host controls the participation of each participant in the program. Part of the acting includes the hosts' rehearsal of the careful timing between talk, dramatic tension, and routine interruptions which the format of the show imposes. More specifically, each talk show is built around several segments which transpose us from the world of drama to yet another type of carnival: the festive images of advertising and pleasure of consumerism.

Further, editing is used in this context not simply in its usual technical sense but in its moralistic dimension as well. More often than not, television show hosts become, with the help of their social psychologists and philanthropists, the arbiters of taste and norms, the defenders of "good" versus evil and the champions of "normalcy"as they are all defined by the official canons in society. These moral dimensions will be fully explored in further chapters. Suffice it to say at this stage that talk hosts play many important roles some of which go well beyond moderating talk or action on the show. They are the embodiment of the forces of "normalization" and the recuperation for all that is deemed "vulgar," inappropriate, or unconventional by the established social order. As will be clearly discussed through the examples of teens and women in spectacle: the host inhabits the space between the acceptable norms of society and the loathed "anomalies" in the system. And very in-line with the ambivalent spirit of the shows themselves, they seduce us by their appeal to liberal democracy, and betray us by their allegiance to the status quo.

In brief, what we have tried to argue so far is that talk shows' carnivalesque appeal lies in their inclusion of a set of ritualized theatrics adapted from various popular forms and genres of entertainment. It is this appeal that vulture culture capitalizes on and utilizes to reposition itself.

Through this process of appropriation, the televised programs continue to reveal a spectacularly eclectic and elastic character. What results from this is the shows' embodiment of several types of contradictions whereby their form becomes often ambivalent and their objectives mostly ambiguous. This is so because eclecticism and elasticity permit the programs to first enhance their entertaining quality and carnivalesque appeal: diversity is an essential condition for building a faithful circle of viewers and sustaining a regular level of interest. Second, they also allow the programs to constantly reinvent and reposition themselves in order to penetrate newer markets and reach larger segments of the public. Third, because talk shows try to reconcile disparate traditions, influences, and genres into a spectacular whole, they inevitably result in a series of troubling contradictions at the level of the knowledges they produce and the ambitions they claim to fulfill.

They pay homage to the established moral and social order, and yet revel in the transgressive forces which threaten to destabilize it. Their response to social crisis and personal dilemmas navigates simultaneously between the poles of conservatism and progressivism. Thus, while practically all shows claim that it is the liberating spirit that animates them, their overall design is largely set to heighten the visual and oral dimension of the spectacle-show. The promise, therefore, to provide a democratic space for public debates comes with a considerably high price: the reinscription of old-time stereotypes and rehearsal of intolerance towards society's "others," as the section below will reveal.

CARNIVAL AND CRISIS: WHENCE THE GEEK AND THE FREAK?

That television talk shows capitalize on the face of human misery is hardly a new argument to make, although this is certainly an important point in itself. What is interesting to explore at this stage is how the daily programs turn serious crises into a carnivalesque spectacle. If the classical definition of newsworthiness is to report not on "a dog biting a man" but on "a man biting a dog," then talk shows invariably adhere to this formula. However, the spectacle of crisis we are offered daily is rife with entertainment quality and political intensity. As previously discussed, temporary "license" of the carnivalesque space is but a first step towards a process of controlling and policing that which is loathed or feared. This is to say that the space of talk shows is heavily charged with ideological meanings and implications most of which ultimately reinstate the dominant vision of how life is to be lived and which specific sets of norms ought to be respected. Talk shows may momentarily humor us with social "absurdities," "oddities," or even human tragedies, but they do so only according to the prevalent social prejudices and stereotypes about gender, age, sex, race, class, and ethnic differences.

It is in this context that we propose to investigate talk shows' enduring fascination with the spectacle of the "geek," and the "freak" turned into the "chic." And it is with the above ideas in mind that this section discusses the shows' spectacularization of two specific social categories: youth and women. In focusing on these two subjects and/or objects of the spectacle, the aim is to: a) illustrate the programs' carnivalesque dimensions which we have discussed in rather general terms so far; and, b) reveal their complicity in "high" cultural norms and expectations despite their appeal to "low" forms of cultural experiences and expressions.

I. THE SPECTACLE OF YOUTH IN CRISIS

How many times have talks shows invited us to indulge in the spectacle of "wild teens," "teen out of control," "wild teens and sex," "my teens are violent," "my teens won't hit the books," "my teens are obsessed with looks," the "teens fashion make-over," or "from geek to chic"? The answer to this question might seem almost irrelevant were it not for the compulsive frequency with which talk shows stage "youth" as the dangerous element which threatens the imagined equilibrium of the social fabric.

The fascination of talk shows with teens is, in fact, a re-enactment as well as reinforcement of the ambivalent attitude which modern society reserves for youth: a mixture of fear and desire. Society simultaneously celebrates youth as pleasure, and condemns it as trouble.[21] Youth is a pleasurable desire because modern society defines its canons of beauty in terms of eternal youth and "ageless-ness." This much has already been argued by scores of communication researchers, cultural analysts, and media critics. Youth-beauty is the driving engine behind the multibillion dollars cosmetics and advertising industries. Youth-beauty is a foundational ideology in the patriarchal definitions of femininity. In brief, youth-beauty is a cult.[22]

Yet, youth is also a source of fear which troubles society's sense of discipline, order, and control. This is so because youth life styles and patterns of behavior do not always conform to society's established codes and norms; nor do they plead allegiance to its aesthetic sense of beauty. When this happens, as it usually does during any process of self-definition and identity-confirmation, the young are considered "outrageous," "little monsters" who are "out of control"(all labels frequently used by talk shows). They are most threatening when their behavior violates the established expectations of the family, school, church, and several other social institutions which consistently mold their identity. Strict measures of discipline and surveillance become, thus, justified in their conception and application. A more detailed discussion of discipline and control of young bodies is offered in chapter five.

However, as Charles Acland aptly argues, monitoring the little "criminals within" is no longer the responsibility of parents, school teachers and counselors: the entertainment industry has long stepped in to add its own reflection on the crisis. And, youth in crisis is a profitable spectacle because it has to do with performance: "it is good magazine and newspaper cover copy, it makes for popular reading and movie-going, and it is the topic for policy debates in public offices."[23] A close analysis of *Montel*'s "Teen Make Over" and *Sally*'s "My Wild Teen Needs Boot Camp," confirms the validity of the above arguments.

The focus of *Montel*'s show aired the first week of December, 1997 is a "dramatic teens makeover" meant to attenuate the parents' embarrassment by their children's "outrageous" dressing style. Right from the opening, the show leaves no doubt as to its own definition of the teens' gender: it is an exclusively female category. The show

takes us first to pre-recorded home scenes of young girls commenting on the "whys" and "hows" of their specific selection of "pecu-
liar" dress styles. Each one of the recorded scenes makes, in fact, a
clear statement about these teenagers' struggle with identity crisis,
confusion, and search for individual recognition. Montel Williams
not only ignores the source or nature of these outcries, but heed-
lessly proceeds to tame the different "outrageous" styles through a
staged dramatic makeover. Fashion experts and beauty specialists are
invited to deploy their skills to either cover or remove the tattoos
from the teens' bodies, then, transform their exotic look into a more
"acceptable" style—read conventional.

While Montel Williams extorts promises from the teens to be
less shocking and boisterous in the future, he proceeds to a dramatic
makeover of their mothers. Here, it was considered fit by the show
host and his experts that the mothers should cultivate a younger
style that should eliminate at least ten years from their mirror image. Thus, while the teens are robbed of their stylistic expression of
youth, their mothers are blessed with the sex-appeal of the ageless
movie stars or soap opera figures.

The last ironic twist in the show, which is significantly the most
dramatic of all, occurs when Montel Williams offers himself to the
expert hands to have a tattoo design 'inscribed' on his shoulder.
With a proud display of his tattooed body, the host keeps reminding
the made-over teens of their need to conform in dress and behavior.
The show ends with interesting comments from the fashion editor
of "Twist" magazine on the latest fads and trends in youth fashion.
Interestingly enough, the cover of the magazine as well as its editor
clearly argue for the importance of encouraging teens to experiment
with different styles and choose their own ways of expressing their
own personality.

Thus, the show has produced in a single episode so many contradictory images and discourses, and constructed as many divergent
bases of knowledge that the implications can only be confusing if not
unsettling. That is, the show has blurred the boundaries between
what is appropriate and what is inadmissible across gender, age, class
and racial lines. The swapping of roles during the entire show simply exacerbates the emotional, cognitive, and psychological confusion that the teens were trying to cope with through their initial

choice of outfits. No doubt, amidst the chaos of contradictory statements, the voices of conservatism and patriarchal order remain the
loudest. This pattern is repeated with a staggering consistency over
a number of talk shows' spectacular "make-over" scenes, irrespective
of the orientation hosts want to give to their individual programs.

Whether we refer to the teens' "outrageous" dress style or their
"out-of-bounds" behavior, we are in essence dealing with an important segment of the population which is in dialogue with larger
social conflicts and deeper contradictions in society's value system.
Youth struggle with social contradictions and identity confusion is
expressed "obliquely," as Dick Hebdige aptly contends, through
their refusal to submit to "forces of normalization" they are exposed
to on a regular basis in the school, family, and through the plethora
of media products.[24] Defiance and discomfort, then, find expression
in broken codes of dressing and living, and in challenging ways of
talking and treating figures of authority. Defiance is communicated, ironically enough, through mass-produced clothing and related
accessories that the consumer market builds for and around teens.
Thus while talk shows exhort teens to tame their rebellious looks,
media outlets—such as teens magazines, television sitcoms, and, music channels, Hollywood movies and cyber malls—urge them to cultivate their individuality through the consumption of "alternative"
images and goods. This is one of the major contradictions youth
register and respond to.

In saying this, we are not disputing the existence or urgency of a
"crisis"; but we are questioning society's misleading tendency to label
it exclusively a "youth" problem. And talk shows are fully implicated
in this process of labeling. They dramatize it, they spectacularize it;
but they do not seriously address the increasing failure of modern
society to respond to its "little criminals" in a knowledgeable and
competent manner. Thus, while the methods of displaying the spectacle of youth might change with the technologies we possess, the
end-goal of staging the crisis is virtually the same: how to preserve
the normative boundaries of socially acceptable behavior.[25]

2. THE SPECTACLE OF WOMEN IN CRISIS

Research on the audience composition of talk shows has already
underlined the considerable visibility of women as guests, calling

participants and audience members of the programs.[26] Since their inception, daytime shows have persistently appealed to the eyes, hearts, and conscience of women, as well as to their narrating skills and emotional constitution. If society traditionally defines its women as an endless stream of talking, complaining, gossiping or "bitching,"[27] then talk shows thrive on these images and canonize them in order to maintain their carnivalesque appeal. This is not to say that male presence or experience is absent from the shows: men's narratives and concerns are just as prized on the daily programs. However, women, as we are often reminded by society and its media, are better story tellers, more sensitive, and more emotionally demonstrative.

Talk shows have, thus, responded to these definitions by depending heavily on women to provide both the raw materials for the daily narratives and the melodramatic dimension for staging them. In other words, women have grown to constitute that essential link between the "talk" and the "show" in the carnivalesque world of daytime talk shows. Images of a sobbing woman relating the injustices of a son or a daughter, and the infidelity of a husband, lover, friend or sister, is more than a familiar sight on these programs. But so is the spectacle of the hysterical woman arguing for her "out-of-bounds" sexuality or enacting her "outrageous" transgression of the established definitions of femininity and womanhood.

The viewers of the *Jerry Springer* show, for instance, are regularly invited to revel in the images of "sex strippers exposed"; "I'm a teen call girl"; "online strippers and escorts." They are also encouraged to actively respond to such narratives as "I have many lovers"; "I'm sleeping with your man"; "It's your bachelor party or me"; or, "Give up your sexy job." The spectacle of Jerry's "freaks" or "geeks" invariably invites hysterical reactions from the public which usually vacillates between fascination and repugnance, as well as laughter and condemnation. This is an example of the carnivalesque laughter discussed earlier.

More "viewer-friendly" shows, such as *Sally, Montel, Maury, Ricki*, or, *Jenny Jones*, might not embrace Jerry's version of the carnival but reveal, nonetheless, an equally keen enthusiasm for the spectacle of a sobbing or hysterical woman. More often than not, these shows revolve around narratives of women who have betrayed through lies, deception or malice, the normative behaviors prescribed by society, family, and other social institutions. In these cases, the perpetrator of injustice exposes herself to the vocal hostility and humiliation from the audience while the victims of her transgression are rewarded with tears of empathy from viewers and gestures of sympathy from the show host.

In other words, the range of roles women are often expected to re-enact on the carnivalesque stage of the shows is fairly narrow in scope and vision. They mostly revolve around performances of woman-as-victim and woman-as-transgressor. The former is the custodian of the morality in society; the latter is the violator of the established norms. Again the first is an image of sacrifice, endurance and misfortune while the second is the embodiment of vice, intolerance, and injustice. To put it crudely, one is a saint; the other is a whore. The shows' reductive polarization of the madonna versus the harlot extends as well as reinforces the long patriarchal tradition of framing women in and around these two central images.[28]

What talk shows offer us on a daily basis is a variation around the same themes, to which are added the melodramatic effects of tears or laughter, and sometimes both. The woman-as-victim frequently denouncing her out-of control teen, for instance, is a re-enactment of the same role of moral custodian who is frustrated in the fulfillment of her duties. The shows' dramatization of her helplessness, and the hosts' gestures of rescue, are all instances which speak directly to the role women are still expected to play in the preservation of the established moral order.

On the other hand, the female victim of betrayal is usually invited to enact her pain or rage on the show by directly confronting her traitors. Whether these tormentors appear in the figure of a female family member, a friend or her partner's lover, the confrontation is staged in a spectacularly dramatic fashion. The woman-as-victim becomes the epitome of victimization, while her traitor emerges as the embodiment of transgression. Here, the audience members and the show hosts often act as the moral judges and witnesses of the enfolding dramas.

It is this persistently bipolar vision of the shows, referred to earlier, which is troubling in its meaning and implications. In this case, polarization not only keeps the old saint/harlot images of women alive, but mostly freezes them in time and presents them as perennial truths. Polarization makes fun of women's real concerns: it builds a spectacle around their confrontation and invites us to watch their hysterical tears and outbursts of anger, their verbal hemorrhage or uncontrollable laughter. This vision of emotional excess with which women are often associated on the shows reconstructs the classical example of the "grotesque" female body.

The bodily displays of woman-as-victim, woman as transgressor are among the most profitable acts of performance which talk shows fetishize and capitalize on. Very much in the tradition of the earlier carnivalesque world, woman becomes a concrete vision of excessiveness, openness, and exuberance: appealing and repelling at the same time. In her fascinating study on the "female grotesque," Mary Russo argues that the female body in the patriarchal social order becomes the locus of strong associations between the grotesque, the outrageous, the hilarious, the excessive, the delirious and the debased.[29] This is so because oppressive patriarchy validates its oppression by perpetrating images of women as a hysteric figure, ungrounded and out of bounds constantly enacting her anguish or rebellion.

The fear of "making a spectacle" of oneself, as the researcher rightly points out, is almost an institutionalized way in which women construct their femininity and relate to their body during different stages of their life.[30] This fear is also closely associated with the notion of shame and embarrassment which an "inappropriate" display of emotion or action might entail. Making a spectacle of oneself establishes, thus, the parameters of femininity and defines the sets of feminine behaviors which are tolerated in the public space. Deviation from the norm exposes the female figure to humiliation and ridicule by the others.

Television talk shows do not challenge the above patriarchal definitions: their spectacle depends too much on the tears and fears of women. They build their carnival on the old tradition of "sobbing sisters" while they reap the benefit of real life dramas and human traumas.[31] Capitalization on and the appeal to women's tears are not exclusively prized by daytime shows, although no other genre of popular entertainment relies on them so systematically or so persistently as talk shows do. Although different media genres share women's tears as their focal point, talk shows have championed them in exploiting the carnivalesque appeal of tears and direct confrontation on stage. Virtually all daytime talk show hosts outperform each other in the amount of tears they succeed in extracting from their victimized or transgressive female guests. The success of the spectacle depends too much on them.

Conclusion

Although the classical model of the carnivalesque space, as revealed in Bakhtin's work, does not address the specific complexities of twentieth- and twenty first-century popular entertainment, it has proved to be a useful category for analyzing the ambiguous nature of television daytime talk shows. Specifically, it has been helpful in demonstrating the close links between cultural politics and the spectacle in popular media culture. More importantly, perhaps, it has allowed us to explore yet another aspect of vulture culture. The carnivalesque dimension of vulture culture reveals how the programs play to both the reactionary/conservative segments of society and cater to its marginalized and transgressive elements. Vulture culture appeals to many entertainment genres and forms of discourse and settles in none for too long. Moreover, it makes a public spectacle of human emotions and private passions but escapes easy classification precisely because this culture consists of hybrid forms and contents.

It should come as no surprise that for some critics and viewers alike, the pleasure derived from watching talk shows easily slips into the dangerous and the toxic. For others, the programs are too crossed by patriarchy and capitalism to constitute any democratic spaces for meaningful social change. Still for others, the shows' ambiguity and ambivalence in both forms and content do not exclude potential spaces for accommodating both pleasure and resistance.

Ultimately, one can easily argue that the "society of spectacle," as Debord rightly refers to the capitalist consumerist order (1990), has produced its electronic version of the carnival: a truncated model that neutralizes the progressive elements of the earlier popular model. The electronic carnival is, therefore, neither a totally liberating space for opinion formation nor an entirely democratic site of information exchange. It confounds the boundaries of familiar genres and settles comfortably in the interstices of double meanings.

6

BODIES DEFINED AND CONFINED

They were the children wise beyond their years, precocious little girls, ambiguous schoolboys, dubious servants and educators, cruel or maniacal husbands, solitary collectors, ramblers with bizarre impulses.[1]

"Precocious" girls, "ambiguous" boys, "cruel" husbands, "bizarre impulses"—and sometimes "cruel husbands" with "bizarre impulses"—this is the stuff of the daytime television talk show.[2] The quote above from Michel Foucault, while unwittingly describing (stereo)typical talk show guests, also leaves out the frequent key players in the daytime television talk show—women and their bodies. Sexy bodies, sexual bodies, motherly bodies, teenage bodies, old bodies, firm bodies, augmented bodies, fat bodies, thin bodies, misformed bodies and beaten bodies—female bodies occupy distinct places in the construction of the daytime television talk show. The daytime television talk show defines and redefines acceptable behaviors and gender norms[3] for women and men, with regards to sexuality, femininity, and masculinity, as well as reform solutions for those women who do not conform to standard notions of womanhood and femininity.

In the previous chapter, we discussed the ways in which narratives, hosts, audiences, and guests are spectacularized in a carnivalesque display. This chapter begins by briefly exploring the ways in which bodies and gender norms are defined on talk shows according to systems, traditions, and social institutions rooted in disciplines and power. By gender norms, we refer to the set of expectations, roles, and behaviors that society constructs/reserves for its women and men. Gender norms define, for example, the specific images and expressions of femaleness and masculinity within society. Any change in gender norms is often perceived as devious and hence resisted, contested, and even contained. An elaboration of the types of episodes typical of the daytime television talk show allows us to further examine the ways in which bodies (sex) and sexuality (gender), both male and female, are presented, defined, challenged, and recreated. Exploring the negotiation of the body and gender norms as it plays out in these episode types reveals the mechanisms of vulture culture at work—the ways in which entertainment television mobilizes expert knowledge, institutional structures, and pre-existing notions of gender to position women's and men's bodies in particular, often oppressive, ways.

POWER AND THE BODY

Implicit in any discussion of gender and bodies are questions of definition and of power. When we use the term power, we recognize that power is a tricky thing to talk about as the word itself has multiple connotations. Briefly, in this instance, we do not envision power as a top-down or bottom-up operation. Power is systematic; it works through the institutions and practices which produce and consume knowledge and that naturalize knowledge and ideologies as truths. In short, they define the body and its relationship to the world. Sandra Lee Bartky expands on the amorphousness of power with regards to femininity. She asserts, "The disciplinary power that inscribes femininity in the female body is everywhere and it is nowhere; the disciplinarian is everyone and yet no one."[4] In this way, particular norms of sexuality and acceptable representations of the body are normalized through such institutions as family, health/welfare and education, rendering them invisible, taken-for-granted commonsense practices.

Michel Foucault, arguably the seminal author on sexuality, discipline and power, writes that understanding the "polymorphous techniques of power," with regard to sexuality, "account for the fact that it is spoken about, to discover who does the speaking, the positions and viewpoints from which they speak, the institutions which prompt people to speak about it and which store and distribute the things that are said."[5] This chapter will explore one specific site (the daytime television talk show), but with the same goals—illuminating the subject, identifying the speakers, the contexts and the consequences—to reveal how an investigation of femininity and sexuality can illuminate a complex system of power and discipline. Understanding power in this way, refocuses questions of discipline and control—suggesting that the ways in which women's bodies are managed are done on multiple levels and in a variety of ways. Discipline and control are both physical and psychological, resulting in both self-control and external control.

Our initial interest in the "body" was sparked by the "teen-out-of-control" talk shows. In these programs, teen and pre-teen girls exhibit behavior that does not conform to gender norms. As a result, within the talk show milieu, there is a goal of reform, an attempt to realign these girls with "acceptable" modes of dress and behavior. This reform is usually resolved on and through the body through makeovers, public humiliation, or increasingly boot camp/jail. One begins to wonder about the logic of such punishment/reform solutions. Where did these solutions come from? Who suggested them? And, why do talk shows use them? These questions pointed us to questions about the institutions and discourses that define what is meant by gender and gender norms. We place the talk show at the intersection of the definitions and discourses of sexuality and the rationalization of discipline. As such, the talk show implicitly becomes a site through which power is employed and deployed.

As the survey of talk show topics below illustrates, women are most often the subject of talk show narratives that feature their bodies and/or gender norms. And, as we will see, the ways in which women are positioned on these shows provides us with the opportunity to see the implications of definition and situation. Below is a table compiling a collection of show titles from three separate weeks of *Maury, Jenny Jones,* and *Ricki Lake.*

Air Date	Maury	Ricki Lake	Jenny Jones
10-01-01	Please Don't Stare, We're Not Monsters[6]	My Rolls are Phat. . .Step Aside, Sister, 'Cuz Your Skinny Butt Ain't Where It's At!	*You're Going Too Far, Pushin' Your Kid To Be A Star*
10-02-01	Shocking Close Calls Caught On Tape	*Mom, You Think You're Mother of the Year. . .Think Again!*	Jenny, It's Bringing Me To Tears, Help Me Overcome My Biggest Fear
10-03-01	Grandma, Stop Dressing So Sexy	*You Say You Shoplift to Support Your Family. . .Get a Real Job or Get Out!*	Double Down, America's Sexiest Twins Are In Town
10-04-01	3 Years Later, They're Back! Shocking Updates![7]	Hey! You're Pretty Smart! Can You Tell Whose Breasts Are Real and Whose Are Fake?	**You're The Reason She's So Wild, Stop Influencing My Child**
10-05-01	Teen Newlyweds With Babies Ask "Are you Cheating?"	*Back Off! I Say My Drinking and Drugging Are Good for My Baby!*	*Quit Acting Out On The Wild Scene, Wake Up, You're No Longer A Teen[8]*
11-05-01	Are They Mothers, Daughters Or Sisters? Can You Tell?	*You are my Baby's Daddy!*	*The DNA Test Will Tell Who Is The Father Of My Baby!*
11-06-01	*I Love You! Stop Destroying My Life!*	One More Date Before Divorce!	*I'll Find Out Today If You're The Father Of My Baby!*

Air Date	Maury	Ricki Lake	Jenny Jones
11–07–01	Little Kids, Big Wishes!	Temptation on Manhattan Island. . . With a Twist!	I Flash My Body, 'Cuz I'm The Next Girls Gone Wild Hottie!
11–08–01	*Today, Shocking Teen Paternity Test Results! Part 1*	Sexy Lingerie Makeovers!	**Shorty, I Don't Want To Be Mean! Those Clothes Are Too Tight To Roll Up On The Scene?**[9]
11–09–01	*Today, Shocking Teen Paternity Test Results! Part 2*	Now That We're Married, my Husband's Gotten Fat!	My Looks Were Tore Up, But Now I'm Sexy From The Floor Up
12–10–01	Wild, Shocking Caught On Tape Moments!	Hook Me Up With A Uniformed Hunk!	What Are You Doin'? You're No Model, So Stop Foolin'
12–11–01	"My Child Is Missing… Please Help Me Find Her!"	*A Woman's Place is in the Home!*	*You're Goin' Too Far, Pushin' Your Kid To Be A Star*
12–12–01	**Shocking Teen Secrets Revealed! Part 1**	*Mom, You Can't Stop Me From Backyard Wresling![sic]*	**My Teen's Boyfriend Is Too Controlling**
12–13–01	**Shocking Teen Secrets Revealed! Part 2**	Ambush Lie Detector Tests![10]	Old School Jenny: Cheating Mate Updates!
12–14–01	Night And Day Opposites & They're Madly In Love!	Ricki's Amazing Couples Race!	*Jenny, Help My Teen Boy Stop The Violence"*

An examination of the table on the previous two pages reveals ten episodes dealing with questions of paternity/fidelity (*these episodes are indicated by this typeface*). There are nine episodes about women's bodies and appearance (these episodes are indicated by this typeface). Four episodes are about teenage girls who display "out-of-control" behavior—most notably promiscuity (**these episodes are indicated by this typeface**), although there is considerable overlap on this one. The shows about teen paternity also have elements of the "out-of-control" teenage girl associated with them. If we include teen paternity shows, there are really seven episodes dealing with the teen/woman "out-of-control." There are six shows focusing on the delinquent/powerless mother (*these episodes are indicated by this typeface*). Again, there is overlap. The teen shows usually implicate the parents as bad and/or powerless parents (and the parents are most often mothers). Finally, the last general category in relation to this study deals with hyper-masculinity. These shows feature controlling and/or violent men. Often these men use threatening language and brute force against their usually female partners. Four of these programs aired during this survey (***these episodes are indicated by this typeface***). While this survey is by no means exhaustive of talk show possibilities, it does show that in fact women—their bodies and related gender norms are the "privileged" subject of the daytime television talk show. It is worth noting that different programs may have greater or less frequency in any given week. However, all of the topic groups listed above are indeed representative of talk show fare. One program type, though, is specifically absent. Shows dealing with questions of sexuality—homosexual, bisexual, transgender, etc.—were not represented in this survey, but are certainly staples of talk shows. These shows also tend to have overlap with other program types—notably appearance and hyper-masculinity.

WOMEN AND TALK SHOWS: BODIES AND GENDER NORMS

A further examination of these programs is in order at this point—one highlighting the discursive and disciplinary functions of the talk show and illuminating *why* this sort of inquiry is important. According to Davis and Mares' empirical study of the effects of talk show viewing, the repetition of talk show topics has measurable consequences: "heavy talk show viewers related issues such as teen pregnancy and homelessness as more problematic than light viewers. If anything, there seems to be evidence of an agenda-setting effect. Frequent depiction of a social problem leads to greater perception of the importance of that issue."[11] And, here lies the tension embodied by the daytime television talk show. It has the power, the influence, to "create" an issue and to make it seem relevant, at least in the eyes of heavy talk show viewers. As such, it is important to take account of the bodies associated with these "issues" of paternity, fidelity, parenting, sex and sexuality, violence and conformity. In many ways, these bodies come to represent these issues—are the physical manifestation of these issues. Bodies are the sites through which these issues are lived and experienced. Here we ask what it might mean for women of a blue-collar or working-class background to participate on a show about paternity, or teen-out-of-control? Does it mean something different if this same woman is also black, Latina, or mixed race? What might it mean for a black woman to have her hair straightened on a make-over show? These sorts of questions necessitate an accounting of the race, class, and sex/gender of these bodies in order to understand their histories and account for their presence and, in some cases, absence.

RACE AND CLASS

While our argument centers on female bodies, these same bodies are also raced bodies and classed bodies. The presence of these bodies as separate, yet one and the same, haunts the talk show and much of the work on them. We know they are there, but do not know how to call them forth. Furthermore, we have to understand these bodies as representations—mediated, constructed, and stereotyped. As representations, their complexity, their unique lived experience is collapsed into a familiar, formulaic "guest," denied a history and a past beyond the confines of the program.[12] We argue that the female body is most often foregrounded, while the raced and classed body haunts the background, silently signaling stereotypical codes. So the question becomes how and in what ways are these stereotypes false, and yet fixed, in our minds and our institutions? Where is the evidence of racially stereotyped bodies?

With regard to the talk show, it is important to understand how sexed, raced, and classed bodies are articulated. Sexed bodies are articulated initially by sight, but also by context; in this case, program topic and the body's position in relation to the topic. It is also through context that sexed bodies become gendered bodies on talk shows. For example, a show titled "That's not a Boy! That's My Daughter" frames a body that at first sight looks male, but then situates and redefines it as a female body through a make-over. Classed bodies, however, are more challenging; they are not as "visible" as sexed or raced bodies. Classed bodies are articulated through language/education, dress, and context—the type of show and the very presence of the body on the show. Talk shows rely on the social and economic positioning of working-class and poor women. Historically, it is not uncommon for these women to appeal to public welfare institutions to resolve family problems or to seek forms of support.[13] Today, talk shows offer these same families (read female-headed households) the "opportunities" to get "help" in a public forum. Whether the appeal is to state-sponsored assistance or talk show-sponsored assistance, *there are always consequences.* In the welfare system, being public means being monitored and labeled as poor, dependent, and needy in order to have access to the resources. On talk shows, being public means carrying the same labels and being the spectacle. However, we should not assume that all guests on the show are working-class. On the "That's Not a Boy; That's My Daughter" episode mentioned above, the mothers and children are decidedly middle-class. During that episode, one grandmother talks of offering her granddaughter $100 to wear a dress. Another mother is a fashion designer. And, all of the girls were involved in organized sports. While organized athletics does not always indicate middle class, we can easily imagine these particular mothers as "soccer moms." Through class indicators such as fashion, language, and education, we can begin to see how female bodies are also classed bodies.

Raced bodies, on the other hand, are articulated by skin color initially. But, they are also articulated by class, and so by their very presence on the show. Given the correlation, both real and imagined, between minorities and poverty, the connection between race and class is an intimate one. Stuart Hall argues, "Race is thus, also, the modality in which class is 'lived,' the medium through which class relations are experienced, the form in which it is appropriated and 'fought through.'"[14] With regard to female, raced bodies, bell hooks reminds us that "[t]he woman who is seen as inferior because of her sex can also be seen as superior because of her race."[15] The sexed/raced body has opportunities and limitations that are defined by sex/gender and by race. Examples of this are articulated in Ricki Solinger's *Wake Up Little Suzie* where she shows the ways in which white single mothers are seen as having a psychological problem whereas black single mothers are seen as social dregs and "wanton breeders."[16] But these observations about race do not necessarily play out as clearly on the talk show stage. Talk show topics rarely focus on overt issues of race, or class for that matter—this is precisely the point we're making in this chapter. And yet, guests represent a variety of races. Knowing that race and class are crucial to any discussions and understandings of talk shows, any analyses would be deficient without being able to take account of their implications.

So it becomes a matter of finding the right questions to get past the absence of an overtly identified raced body—one that comes to the show along with its histories of colonization, racism, domination, and exclusion. How does the talk show account for raced and classed bodies? The answer is that it doesn't. The talk show discourages thinking about raced and classed bodies. The inability to think about race and class with regards to the talk show IS the problem with race and class on talk shows. Evidence of race and class discrimination in the program topic is present by its very absence. The sexed, raced *and* classed body has been staring at us the entire time, but has never been allowed to speak out its history. The presence of raced and classed bodies remains unseen because these bodies are never articulated as such. But, this is more than just a question of articulation of particular bodies. What is revealed here is the construction of spaces which disavow the presence, legitimacy, and recognition of particular bodies, experiences, and histories.[17] This erasure or disavowal is precisely the reason we have to understand vulture culture. Vulture culture intrigues us not only by what it allows to be said, but also by what it prefers to silence and/or erase.

FEMALE BODIES THROUGH THE MULTICUTURAL LENS

Talk shows are structured by discourses about race, class, and sexuality couched in liberalism and multiculturalism. These discourses, not unique to the talk show, set the terrain of struggle and write the script through which these discourses are acted out (both on the talk show stage and in our everyday lives). Female bodies and norms are addressed in virtually every episode, while raced and classed bodies haunt the script. Their presence is rarely acknowledged.

This disavowal of the raced and classed bodies on talk shows works with and through liberal discourses of multiculturalism. Once held as a radical possibility, there were high hopes for multiculturalism as a solution for racial equity in the late 1980s and early 1990s. Scholars, educators, and policy makers believed multiculturalism could challenge the neutrality of whiteness and as such, challenge race, sex, and class privileges inherent in dominant social institutions such as education, the justice system, employment services, and welfare agencies. Sociologists Christopher Newfield and Avery Gordon argue that multiculturalism must have a political component: "A real multiculturalism requires political as well as cultural inclusion, requires the sharing of power among relevant groups. . . cultural diversity within institutions that are still white-dominated lacks redeeming social value."[18] Similarly, legal and feminist scholar Angela Davis argues, "A multiculturalism that does not acknowledge the political character of culture will not, I am sure, lead toward the dismantling of racist, sexist, homophobic, economically exploitative institutions."[19] These revolutionary hopes for change were short lived.

But, the realization and implementation of multicultural discourses have been filtered through liberalism and individualism. Legal scholar Wendy Brown points to the tension between universal representation and individualism:

> . . .the latent conflict in liberalism between universal representation and individualism remains latent, remains unpoliticized, as long as stratifying powers in civil society remain naturalized, as long as the "I" remains politically unarticulated, as long as it is willing to have its freedom represented abstractly, in effect, to subordinate its "I-ness" to the abstract "we" represented by the universal community of the state.[20]

According to Brown, raced, classed, and sexed bodies are articulated merely as human[21] bodies in liberal discourse. Multiculturalism has been appropriated into scripts of liberalism and individualism— indeed, it never existed outside these scripts.

So, what is the legacy of multiculturalism? How has this logic informed the talk show? How is it that the raced and classed bodies rarely get to speak, or to be called into being? We suggest that the talk show has taken up "the culturalism of multiculturalism,"[22] has embraced it even. That is to say, this embracing of multiculturalism erases the material and lived experiences of race. Instead, it views racism as cultural experiences rather than as experiences structured by social, political, and economic institutions. Talk shows frame their topics as if all races and classes might be affected by them in the same degree and with the same frequency. Our argument contends that female bodies and gendered norms are the subject of talk shows. But, these are also raced and classed bodies with historically different experiences. Talk shows reduce these bodies to female bodies, ignoring the implications of race, but relying on notions of class. Talk show host Montel Williams explains his show's take on the influence of race and class on one's social, political, and economic circumstances:

> "Oh, this happened to me because my father beat me." "This happened to me because we were poor." "This happened to me because I'm black." Well, woe is me. Face up to it people: if you're in the minority in this country, you will always feel like an outcast. Your children will always feel like outcasts. You'll have it tougher than most, most of the time. That's the way it's been, that's the way it is, that's the way it will be. Deal with it. I have long ago reconciled myself to the fact that some people will never accept me simply because of the color of my skin. And do you know what? I don't care. I truly don't. Accepted or not, I can still do what I want to do, whatever I'm qualified to do, whatever I'm entitled to do. I can supply my own validation, and I do that by setting a path for accomplishment and by assuming responsibility for staying on course.
>
> Now, I'm not blind to racism. It's out there—still as much as ever— but it doesn't mean we can check out and quit trying. The truth of the matter is racism is no longer the obstacle it once was, and in my book it is certainly no longer an acceptable excuse for any of the failures plaguing some of our young kids today. I'm sorry, I just don't buy the way kids are so quick to pin their troubles on their parents or on the color of the skin. Yes I can see that you're black. I can see that you've got it tough, that no one's ever cut your father a break. I feel for you. I really do. But you can do better. In fact it's your responsibility to do better. It's your job to take your rear end off the streets and sit it down in school or in a library, where it'll do you some good.[23]

Williams, himself a black man, acknowledges racism exists, but it is unclear whether he understands racism to be individualist or structural, whether it is a belief in a person's mind or a discursive and disciplinary construction with implications for social, economic and political institutions. In either case, Williams does make the ties between race and class by nodding to the African-American father who never got a break. But, in saying "racism is no longer the obstacle it was," Williams does not seem to account for the historical positioning of race and poverty. Racism might not have the same dimensions as it did in times past—for instance, it might be as overt as "white-only" drinking fountains—but that is a function of historical change and discourse. To minimize the presence and effects of racism in today's society, in fact, works to perpetuate it. According to Davis, "Obviously, there is a reason we are stuck with the term 'race,' and that has to do with the persistence of racism and processes of racialization that perpetuate race-based oppression, even as they appear to move beyond it."[24]

Williams suggests knowledge and education will pave the road to a better future for black children. Certainly, a laudable message, but it does ignore the fact that children living in poor communities do not have the same educational resources as children living in middle-class communities.[25] Furthermore, education in and of itself is not free from racism.[26] Williams is bound up in the tensions of racism and liberalism. He knows it is there, but believes in the power of the individual to overcome it. He resolves this issue by not focusing on racism. As such, he frames issues and topics on his program outside of race. Ultimately this omission masks the historical implications of race and racism and instead situates the problems as sex/gender issues.

But, the raced body still haunts the talk show. Heaton and Wilson, speaking as psychologist and counselor, suggest, "Since most of the shows pivot around issues related to sex, violence, and relationships, the shows tend to get people of color who reflect stereotypical images. They bring on guests who have illegitimate children, are supported by welfare, fight viciously, and have complex and unsolvable problems."[27] Our experience suggests Heaton and Wilson overstate the matter. In fact, panels are racially mixed (roughly 30–50 percent white), and the women presented make it very clear they are not on welfare. But, Heaton and Wilson do make a valid point. Regardless of the "facts," these are raced bodies that carry with them particular stereotypes. What Heaton and Wilson leave out is drug use. In an episode of *Sally*, Sally Jesse Raphael is asking a Native American girl named Woesha about her use of marijuana. Woesha replies, "Everyone on the reservation does it." Raphael yells back, "I'm sure all the people on the reservations do pot. I'm sure all the people in the projects do pot. I'm sure all the people in the barrio do pot."[28] Raphael's comment clearly associates drug use with minorities and sites of poverty. Notice, she did not say all the people in the suburbs are doing pot. Drugs are seen as a vice of minorities, specifically poor minorities who live on reservations, in the projects, and in barrios. Second, it exacerbates issues of class and poverty, which is where the conservative discourses come into play.

Despite their appeal to liberal multiculturalism, talk shows indeed exploit a wide range of racial and gender stereotypes, with positive and negative consequences—increased visibility framed in stereotypical images. One must wonder, jumping off from Davis and Mares' empirical conclusions, that if heavy talk show viewers have a greater sense of a "social problem" due to frequent repetition of particular topics, what might the show repetitions do for issues and perceptions of race and class?

To better answer these questions, we need to look closely at specific examples taken from the table presented at the beginning of this chapter. These examples will demonstrate the centrality of the female body around which race, class, and gender are articulated. A look at the show types highlighted above will help to illuminate tensions of race and class with regards to gender under the umbrellas of discipline and discourse. While each of the following program themes could certainly be expanded into a full-length chapter, we will provide a brief introduction to these themes and their implications for female/raced/classed bodies, femininity and sexuality in order to situate the female body as the subject of the daytime television talk show. We will then specifically explore the intersections of knowledge, discourse, and female bodies on the talk show by taking an in-depth look at the "teen-out-of-control" show and the societal treatment of deviant teenage girls and their bodies.

APPEARANCE – THAT'S NOT A BOY! THAT'S MY DAUGHTER

Perhaps the most straightforward of all the talk shows, the appearance show places the female body front and center. The goal of this show is to present and reinforce standards of feminine appearance, and to conform female guests to that standard. Sandra Lee Bartky notes the trend of femininity becoming increasingly focused on female bodies—both in appearance and sexuality.[29] We can clearly see this play out in the appearance shows. Of the nine episodes dealing with women's appearance, four of them have the word "sexy" in the title while three others explicitly deal with the definition and recognition of a "sexy" body. These programs can be classified further into show-off episodes, which display the female body without reform measures, and makeover shows, which end in the abnormal body being given a makeover. We should also note that male bodies are not excluded, but other standards are attached to them.

The show-off show varies in sexual intensity, but almost always features a woman's body as a sexual body. While it may not offer reform, it encourages audience members to scrutinize the guests' body to find flaws or lack of "authenticity." For example, *Ricki Lake*'s "Hey! You're Pretty Smart! Can You Tell Whose Breasts Are Real and Whose Are Fake?" features a display of women of "all shapes and sizes" with the goal of determining whose breasts are "home grown" or "not her own."[30] The spectacle of the female body, specifically the breast, becomes tied up with questions of authenticity, body image, and the peep-show.

However, certain types of the show-off show blend with the makeover show. Episodes that present people formerly teased for their looks who now are sexy and attractive are called "geek-to-chic" shows. In these programs, the guest has undergone a radical physical transformation in recent years and she wants to show herself to the people who used to tease her. The guest transforms herself to fit with her perceptions of beauty and sexuality. An example of this type of program would be *Jenny Jones*' "My Looks Were Tore Up, But Now I'm Sexy From the Floor Up." Featuring guests who have not only radically changed their appearance, they have also become exotic dancers. This episode offers us insight not only into notions of how femininity, sexuality, and desire are constructed, but also into notions of class structured around occupational status. Exotic dancers are at the pinnacle of spectacle and female sexuality. Desire through sexuality is used to reaffirm the sexy as attractive. So that, the way to move beyond "geek" status, is to become sexy. However, this transformation comes with a cost. While exotic dancers have spectacle value on the talk show, they do not have much status in society at large. While they may earn "good money" dancing, they do not adhere to middle- and upper-class values of modesty and propriety. They are viewed as déclassé.

Not surprisingly, both types of programs typically present extremes—either a woman is dressing too sexy or not sexy enough. The abundance of these programs is illustrative of the genre's reliance on coded bodily representations of femininity and sexuality. For instance, *Ricki Lake*'s "Sexy Lingerie Makeovers" show synopsis states, "According to the men on today's 'Ricki,' what began as a passionate relationship between their partners and them has since grown cold because their 'better halves' have gone from hitting the sack in sexy sleepwear to donning something that more closely resembles a sack."[31] The *Ricki Lake* solution is transform these women by removing them from their comfortable sleepwear and put them into "sexy" lingerie. What perhaps remains unstated in this show is the expectation that women need to sacrifice in order to fulfill men's desire. So, that guests can "rekindle the 'lust connection' with eye-popping lingerie makeovers guaranteed to fuel the fire of love"[32] suggests that the transformation of the female body into a sexual body is done in order to satisfy the male partner. It is worth noting, though, that Bartky's observations about sexuality and appearance (the sexualization of the female body) do not apply to grandmothers or teen girls in the talk show. The cultural codes of the aging body and the innocent body suggest that when grandmothers and teen girls are sexualized, these bodies need to be rearticulated, reconstructed, according to expected gender norms for grandmothers and girls. In short, they need to be desexualized.

Appearance shows dealing with teenagers almost always involve makeovers, not for sexual appeal but for conformity to age-appropriate ideals of femininity. The program titled "That's Not a Boy! That's My Daughter" works to rearticulate the female teenage body into a properly feminine manner of dressing. The girls on this episode reject traditionally feminine clothing and technologies (makeup, hairstyling products, etc.) The episode's synopsis states, "The moms here today are sick and tired of their daughters dressing and acting like boys!...MAURY's makeover team takes action and turns these tomboy teens into sweet sensations."[33] Mothers, makeover teams, and Maury Povich himself, all work to transform these young female bodies into feminine bodies, but not sexual bodies. And, in all of the cases discussed above, what need to be highlighted are the commonsense assumptions about femininity and sexuality. Adult women need to be sexy for their men, while teen girls are transformed into "sweet sensations." The female body needs to be articulated in the discourses of femininity for it to seem normal, for it to make sense.

The talk show's presentation of women's bodies emerges out of the discourses of fashion, beauty, popular culture, and the sex industry. Future research on this particular type of program could intimately explore the ways these discourses inform and are interpreted by the talk show genre. And, let's not forget that the ways these same discourses construct beauty/sexiness have implications for raced and classed bodies. For instance, in the case of black women, straightened hair is considered more desirable than natural hair. This type of talk show is the one that focuses on the appearance of body—its entire emphasis is on the celebration, critique or manipulation of the body. What needs to be kept in mind when exploring appearance shows is the histories of these bodies—their constructions and their contexts. An increasing and emerging trend on talk shows is to sexualize the male body in similar ways, especially on geek-to-chic programs. An examination of the sexualization of the male body might offer insight into the systems of power which work to position both male and female bodies as sexual bodies.

MOTHERS – "MOM, YOU THINK YOU'RE MOTHER OF THE YEAR . . . THINK AGAIN!"

These types of programs feature guests who have been brought to the show because they are deemed to be unfit or powerless mothers. They are either overprotective of their children or they are considered negligent in tending to their children's needs. In either case, we understand these women as "bad" mothers through a variety of institutional and social discourses. The way in which mothers in our society are positioned is negotiated through the disciplines of education, social work, psychology, medicine, and biology. As such this positioning becomes discourse and is rendered knowledge. These disciplinary and discursive constructions of the female as mother become commonsense, taken-for-granted assumptions about one of the roles of women in our culture. As such, the talk show plays off these assumptions through the display and definition of the "unfit" mother. The program in our survey most intimately tied up with notions of "unfit" mother, gender norms, and the female body is *Ricki Lake*'s "Back Off! I Say My Drinking and Drugging Are Good for My Baby!" The show's synopsis reads, "Pregnant and partying or bearing babies and boozing. . .not the picture-perfect description of a prenatal lifestyle, but the unrepentant moms-to-be on today's 'Ricki' don't think their 24–7 'funfests' are anyone's business but their own."[34] Here the mothers-to-be are trying to stake a claim to their own bodies, no matter how destructive a form that claim takes. However, in our society, woman-as-mother no longer has individual and sole right to her body, especially the pregnant mother. "Proper" mothers are intimately bound to their children—as validated by institutional, disciplinary discourses, especially psychology. As such, the woman's body is put in service of others, be it her children or, as we saw above in our discussion of "Sexy Lingerie Makeovers," her sexual partner. When the woman-as-mother refuses to sublimate her own interests to that of her family, she is considered abnormal.

In terms of race and class, initial observations show the unfit or powerless mother is not predominantly one race more than another. She is however always working-class. Usually, she either works several jobs, or she does not work at all. And, she is almost always a single parent.[35] These women are positioned as unfit or powerless mothers in part because their economic status limits their ability to meet their financial needs, their personal needs, and the needs of the children. Furthermore, according to Davis and Statz, "Although women are often blamed as being the 'cause' of social malaise (e.g., illegitimacy is linked to the mother, not the father), such scapegoating merely sidesteps the real issue—the state's refusal to confront its own failure to reduce crime, domestic violence, child abuse, drug trade, economic inequality, and so on."[36] Furthermore, an understanding of the ways in which women's bodies and their opportunities to mother have been variable over time are impacted by race. For instance, Ricki Solinger discusses the racialized interpretations and perceptions of white and black unwed mothers in the 1950s and

1960s. She argues, "As white unwed mothers were portrayed as a threat to the moral integrity of the family, black unwed mothers were often construed as an economic threat to the same white family."[37] An extended study of the ways in which talk shows present "mothers" might make the connection between conceptions of mothers in political, economic, scientific, and social discourses, on the one hand, their manifestation in popular culture, on the other.

WHERE ARE THE BOYS? THE MALE BODY AND MASCULINITY

As we have seen in the discussion of appearance and mothers, men are by no means absent from the talk show, and we don't wish to oversimplify their presence. It should be understood that constructions of femininity are implicit in constructions of masculinity and vice versa. Heaton and Wilson suggest in their book *Tuning in Trouble* that "[w]omen viewers are given a constant supply of the worst images of men, all the way from rapists and murderers to garden variety liars, cheats, con artists and lazy good-for-nothings."[38]

In our opinion, Heaton and Wilson are close to the mark, but we would organize the categories a bit differently. We suggest that men on talk shows can usually be put into three categories. First, men become the object of the discourse if they are feminized. The most obvious example of this is the gay male or drag queen.[39] Second, men are presented on talk shows as a form of patriarchy. Hosts such as Maury Povich, Montel Williams, Jerry Springer, and more recently Dr. Phil are the symbols of authority, the spokesmen for patriarchy. Not only do they seem to have all the right answers, but they also have the last word. Experts are also used in this way. Male or female, the expert possesses knowledge and/or technical skills, as in the case of lie detector and DNA tests, which work towards the maintenance of the status quo. Also, occasionally, fathers make appearances on talk shows standing in for patriarchal power. Finally, the drill sergeant in the teen-out-of-control programs straddles the line between a "firm" patriarch and the hypermasculine male, which is the third way in which men are portrayed on talk shows. However, we argue that the use of the hyper-masculine man often better illuminates and designates notions of femininity and questions of the female body. We will explore this further with the drill sergeant on the teen-out-of-control programs.

With this understanding of masculinity on talk shows, we want to situate the paternity/fidelity programs and the violent masculinity programs as participants in a dialectical relationship between notions of masculinity and male bodes with notions of femininity and female bodies. Beginning with paternity/fidelity shows, these shows are ones that resolve questions of fatherhood, and have dramatically increased in the past three years as the accessibility to and affordability of DNA testing and polygraph exams has increased. In these instances, the certainty of science is incorporated into the talk show spectacle. And, the spectacle here is to question and/or expose the deviant sexuality of the mother. Shows like "I'll Find out Today If You're the Father of My Baby!" and "Today, Shocking Teen Paternity Results" imply female promiscuity. The words "if" and "shocking" point to moments of uncertainty and doubt. Paternity shows, while attempting to determine fatherhood, also act as a policing of female sexual behavior. For, as mentioned above, mothers are culturally and biologically linked to the children in a way that fathers are not. Women physically bear the consequences for sexual promiscuity when their encounters result in pregnancy. Men do not. There are implications for the raced and classed female body in these shows. Black female bodies have historically been "portrayed by politicians, sociologists, and others in the postwar period as unrestrained, wanton breeders, on the one hand, or as calculating breeders for profit on the other."[40] As such, images of unwed mothers in paternity cases carry this political and cultural baggage.

While DNA testing provides a woman with a legal recourse for support against the father of her child, that recourse takes place in the justice system and not on the talk show stage. And, in order to get the DNA proof she needs for a legal child support claim, the mother must make public her sexuality, her sexual behaviors. The entire show hinges not on the man taking responsibility, but rather proving that the woman is telling the truth, either verifying or denying her claims. Increasingly, the spectacle of the program is intimately tied up with the demonization of the mother as more and more female guests find that the man they claim with all certainty was the father is "excluded" as a DNA match. It is perhaps these moments where the woman is exposed that have made paternity programs a dominant staple of the talk show. The paternity show will be examined in a later chapter in this book—offering useful insights into science, sexuality, technologies, and feminist notions of public and private.

Programs dealing with hypermasculinity and violent masculinity do not take up a major portion of talk show programming schedules, but their presence does need to be considered. To make these concepts clearer, we see violent masculinity as a subset of hyper-masculinity. First, let's examine hypermasculinity. We would generalize the hypermasculine man as being sexist, macho, sometime violent, usually loud, insensitive, and impatient. He can perhaps best be illustrated in a *Jenny Jones* episode featuring a man who wouldn't marry his fiancée because he thinks she is too fat. The woman is attractive and actually thinner than when she and her fiancé first met. The man is portrayed as a stereotypical chauvinist pig. While most people may view this man as the antagonist, we have to keep in mind one of his functions, albeit not a deliberate one. This hypermasculine man works to legitimate and naturalize the emphasis on women's bodies. This episode is about men who wouldn't marry women because of the shape of their bodies. From a functionalist standpoint, the emphasis, the focus, the crux of this entire show is the women's bodies. Here we see the interplay and the mutual dependence of notions of masculinity and femininity. While some of us may think these men are complete idiots, their presence provides the entryway for an emphasis on women, body image, and sexuality.

Violent masculinity most clearly places the man as the villain. These programs feature physically, verbally, and emotionally abusive men who physically control the lives of the women with whom they are involved. While the other programs we have discussed emphasize and manage women's bodies through narrative tropes (makeovers and paternity tests) or cultural constructions (mothers, teenagers), the programs on violent masculinity focus on control by physical force and fear. And, while the talk show rhetorically rejects this form of bodily control, the presence of violent men with women as their victims work to legitimate once again women as the subject of television talk. Furthermore, the presence of hyper-masculinity and violent masculinity may also work to normalize a notion of benevolent men (embodied by the patriarch men such as hosts Montel Williams and Maury Povich), thereby reinforcing structural gender differences and economic inequalities by erasing them. Further research could draw out the differences and similarities between hypermasculinity and benevolent masculinity in relations to the talk show.

TEEN OUT OF CONTROL!

The teen-out-of-control (tooc) programs provide one of the most fruitful sites for the study of ways in which women's/girls' bodies and gender norms are defined and managed in the daytime television talk show. These programs combine elements of each of the program types listed and, as such, offer an opportunity to explore the ways through which women's bodies are incorporated, rationalized, defined, and (re)presented in the talk show. The teen-out-of-control is a site where discipline and sexuality converge. This portion of the chapter will first provide a summary of the program—who is meant by teen? What is meant by out-of-control? What are the punishment/reform measures taken to alter the out-of-control behavior? As we will see, answers to these questions can be found in the logic of punishment and reform which links the talk show to the justice/welfare system. An exploration of how women are treated in both arenas will suggest systematic, disciplinary, and discursive constructions of women, their bodies and their sexuality, that pervade many of our cultural institutions.

So, let's begin by looking at the teen-out-of-control. In any given week on any given day, one is likely to stumble across a show with titles such as *Maury*'s "Help Me Control My Wild, Sexy 13 Year Old," *Ricki Lake*'s "Out of Control Kids Set Straight" or *Jenny Jones*' "My Wild Pre-Teen Needs a Lesson, It's Time for a Boot Camp Session."[41] On these programs you can "meet a twelve-year-old who gets so drunk, she blacks out"[42] or hear the story of "five desperate mothers" and see the talk show aid them in trying "to tame their daughters' wild, violent ways."[43]

A recent two-part *Maury* episode is described on the show's Web site and will serve to give us a flavor of narratives typical of the teen-out-of-control programs:

> The young teen girls here today are violent, vulgar and completely out of control. Fifteen year old Jaycee broke a bottle over another girl's head just

because she was looking at her boyfriend. She is currently on probation for robbery. Fifteen year old Amber has physically attacked her mother. Amber tells her mother today that even though she's in drug rehab, she's still doing drugs. Fourteen year old Silky lives with her aunt and has threatened to kill her by poisoning her food. Silky shocks her aunt today by revealing she has actually put bleach in her aunt's dinner. Fourteen year old Whitney steals cars, sells drugs and is dating a drug addict. Whitney reveals that she has had sex with 15 men. These mothers have no idea what to do with their daughters and think that boot camp is the only option before their young teens end up in jail or dead. MAURY agrees and has arranged a two-day trip to the Labette Correctional Conservation Camp. Plus, Mad Dog takes them to a cemetery and a funeral home to show them the hard, cold reality of death. Will a taste of their own medicine tame these out of control teens?

    The four wild, out of control teen girls on the show yesterday were defiant and disrespectful to their mothers, to MAURY and to the boot camp instructors. They had no fear of Mad Dog or of spending time at boot camp. The teens are back from a two-day stint at boot camp and we are going to find out if it had any effect on their behavior. Amber promises she is going to turn her life around and her mother says she has already seen a difference. Jaycee was probably the most disrespectful of all the girls and thought going to boot camp was a big joke. Her laughter quickly turned to tears and she tells MAURY that boot camp was the hardest thing she's ever had to go through. She says that the first two things she is going to do are to clean up her language and stop physically abusing people. When Whitney saw how she acted on the show on tape, she says she realized how stupid she looked. She says that two days at boot camp definitely taught her a lesson—to respect herself. She says she is trying very hard to treat her family better but admits it's hard to change. Silky says boot camp made her want to turn her life around. Everyone agrees that the girls are on the right track and MAURY will keep up with their progress.[44]

These types of programs focus on the "teen-out-of-control" (tooc). However, as we can see above, "teen" is portrayed in a very particular way. Teen is not a gender-neutral term. In the daytime television talk show, the tooc is female. What the show descriptions imply and what becomes apparent during the shows themselves is that an "out of control" girl is characterized primarily by sexual behaviors, profane language, and disrespect towards authority.[45] And, as is the case of the *Maury* episode above, she may also be violent and/or use drugs or alcohol. Sally Jesse Raphael, talk show host and self-professed pioneer of the tooc program has said, "I've been talking to so-called 'bad girls' for 15 years now. . .[they] hit their mothers, use drugs and are very sexually active."[46] As is obvious in the titles above, the goals of these types of programs is to both punish and reform the girl out-of-control. Interestingly, this reform takes place through and on the body. Depending on the degree of the out-of-control behavior, the girls' punishment/reforms can range from a make-over to public humiliation[47] to a day in boot camp or jail. In the show titled "Help Me Control My Wild, Sexy 13 Year Old," the punishment/reform solution is to make the girls spend some time in an actual prison cell. The *Sally* episode mentioned above is a bit more straightforward— "Send My Wild Teen to Boot Camp!" *Jenny Jones* recruits the help of a drill instructor to send her "guests" to boot camp. And, *Ricki Lake* "enlists the experts for some on-the-spot counseling."

Initially, we were curious as to why the boot camp was seen as a logical form of punishment, why the social workers and counselors on the show allowed children as young as seven to be subjected to stereotypically in-your-face drill sergeants—usually large, loud, intrusive men, occasionally large, loud intrusive women. The talk show's reliance on "expert" knowledge provides us with an interesting entryway into a new understanding and examination of the daytime television talk show—one exploring the ways in which normal and deviant behaviors for women are defined and constructed in talk show narratives. These talk show narratives both inform and are informed by popular discourses coming out of criminology, psychology, social work, fashion, and other forms of popular culture. Furthermore, these discourses have a history of defining women in terms of their sexuality and offer punishment/reform solutions that focus on the body. Yet, an analysis is not complete unless one recognizes the ways in which these discourses work with the talk show's own discourses and conventions. The remainder of this chapter explores women in the justice/welfare systems. This offers insight into the ways in which the deviant woman/girl has been incorporated into institutions in charge of managing and maintaining her and how women become subject/objects of these institutions. As we will see, there are several parallels between women in the justice/welfare system and women on talk shows.

JUSTICE BY GENDER?

We believe that television programs act as a bridge between dominant ideologies (in this case, about what is considered deviant behavior for women and how it should be punished/reformed) and audience members who interpret these ideologies in light of their own life-experiences. In this way, we agree with Todd Gitlin when he writes, "Commercial culture does not *manufacture* ideology; it *relays* and *reproduces* and *processes* and *packages* and *focuses* ideology that is constantly arising from both social elites and from active social groups and movements throughout the society."[48] Here we want to explore this process a bit more, to address the connections between professional discourses and the talk show. Furthermore, we argue a crucial link between the two is the show expert, and a perfect example of this is *Sally*'s expert-in-residence Pat Ferrari.

The *Sally* Web site said of Ferrari, "Boot Camp expert Pat Ferrari has compiled a lot of information for you parents who are considering a boot camp for your child. Pat is a degreed social worker who has had special training in crisis intervention, drug and alcohol care, domestic violence, and rape aftercare (among other things). She is the founder of the Women's Task Force and also founded a battered woman's shelter."[49] We should also mention that Ferrari had her own do-it-at-home boot camp for out-of-control kids that parents could download off the Web for a fee.

Ferrari becomes a bridge between professional knowledge and popular culture as well as a representative of discipline and authority. As mentioned above, several parallels exist between such professional knowledges and women on the talk show. The first of these is the fact that women are punished/censured more often for moral offenses than they are for criminal ones. So, the question, with regards to the daytime television talk show, and the question that we've been working to substantiate throughout this chapter, is why are women consistently the object of the talk show narrative? By examining some of the academic literature concerning women and the justice/welfare system, we may better understand the ways women are positioned in society, especially "deviant" women. In this way, the logic of a talk show's selection of topics and guests can be seen in a larger social context.

An overwhelming amount of research on women in the justice/welfare literature surveyed finds that girls are more often punished for status (read moral) offenses than are boys. Meda Chesney-Lind, a leading scholar on women in the justice system, examines the role of the criminal justice system in the maintenance of patriarchy. Her article "Women and Crime: The Female Offender" offers substantial evidence that women are most often convicted of petty crimes and status offenses. Her research illuminates the fact that women are primarily arrested for minor property crimes, forgery, and prostitution and rarely for serious property crimes and violent crimes "parallels their assigned roles in straight society."[50] However, she also says that no one understands why this is the case. Furthermore, she offers evidence that women who are convicted of crimes are mostly women living in poverty.[51]

Chesney-Lind's research may point to the reasons that the talk show's "teen-out-of-control" is usually female. On the one hand, these girls are usually violent (read abnormal). Violence holds a greater spectacle value. And, this may in return warrant the harsher punishment used in talk shows. On the other hand, the "tooc" may usually be female because she can be controlled, or there is a greater likelihood of control. Chesney-Lind concludes saying "both the construction of women's defiance and society's response to it are colored by women's status as male sexual property. Once a female offender is apprehended, her behavior is scrutinized for evidence that she is beyond the control of patriarchy and if this can be found she is harshly punished."[52] This positioning of women in the justice system can be broken down into three overlapping areas. First, we can begin to see the connection between social problems and the individual—more specifically how the individual becomes the focal point for social reform. Second, we can see the role of self-control. Female offenders have been unable or unwilling to police their bodies—"she is beyond the control of patriarchy." Third, this lack of self-control has resulted in bodily control. Scrutinizing her behavior has meant, as recently as 1980, subjecting women to vaginal examinations and pregnancy tests no matter what the charge brought against her.

THE INDIVIDUALIZATION OF SOCIAL PROBLEMS

The connection between women in the justice and welfare system and morality is played out in the realm of policy and on the talk show stage. This is done in part by aligning "deviant" women with particular social issues. *Moral Panics* by Kenneth Thompson offers the insight that "politicians may find it easier to focus attention on moral issues than to come up with solutions to some more intractable problems, such as lack of education and skills, unemployment, housing conditions, crime and poverty."[53] In this way, moral issues become a distraction and focal point. If we can transform issues of education, crime and poverty into moral issues, the problems are no longer social problems, but individual problems. Hence, punishment/reform can be extracted on an individual and the problem is contained.[54]

A specific example of how teenage girls experience this shift from social problem to individual problem can be seen in Kerry Carrington's ethnographic study on female delinquency. She explores a case from her research where a teenage girl has repeatedly run away from home, has been orphaned, has been sexually abused by relatives in both homes where she stayed, has admitted to being sexually active, and has attempted suicide. The psychologist's assessment of the girl was that she had an inability to cope in the community and, *for her own safety*, ought to be institutionalized. In Carrington's analysis of the case material she concludes, "By making Judy appear individually responsible for the social and familial world in which she inhabited, the rationale provided for punitive intervention masked as benign expertise."[55] This individual emphasis on Judy's problems echoes Thompson's insights into moral panics. Social problems become moral problems, which in turn become individual problems. As we will see, a similar process occurs on the talk show.

In terms of social welfare and the talk show, this individualist isolation exposes women to greater surveillance and censure. This is a key point of tension between individualism and the state. In seeking the help of state welfare agencies, or talk shows, the girls' parents/guardians define the girls as deviant, in need of control. Instead of examining the institutional structures that create the girls' "world," professional knowledges and the talk show format seek to reconcile the girls to their (potentially hostile) environments. Nancy Fraser writes, "As a result of these expert redefinitions, the people whose needs are in question are repositioned. They become individual 'cases' rather than members of social groups or participants in political movements. In addition, they are rendered passive, positioned as potential recipients of predefined services rather than as agents involved in interpreting their needs and shaping their life conditions."[56]

The "teen-out-of-control" show is itself a moral issue symptomatic of a larger struggle with patriarchal institutions (such as the juvenile justice system and welfare agencies). However, the inadequacy of these institutions is not a site of inquiry on the talk show. Instead, the focus is on the specific behavior of the girl, devoid of her life-story. Her reasons for running away, or doing drugs, or having sex usually have something to do with abuse in the home. Her body, positioned not only as a female body, but as a raced and classed body, is never taken into account. We need only to look back to Carrington's (1995) ethnography where teen-out-of-control Judy was repeatedly sent back to a home where she was being sexually abused to understand the importance of life-stories in contextualizing teen-out-of-control behaviors.

The value of this ethnography is in reinforcing what this book argues all along: talk shows should not be taken in isolation. They are situated within a much larger context and related to broader social and political institutions. What we see on the shows is an extension and reinforcement of logic existing in other spheres of society. In this particular instance, we can see the ways in which talk shows mirror and scavenge the disciplinary measures of social work, delinquency, and reform. The case of Judy enables us to trace a particular kind of logic used to "protect" and punish deviant young women. Under vulture culture, the influence given to such logic and reform measures, as part of an institutionalized authority, is unquestioned and normalized.

Like Judy, the talk show teen-out-of control goes through various forms of punishment only to return to the same environment where the abuse continues. The talk show rarely examines the roles of institutions or the life-story of the girl in the "teen-out-of-control" show. This is, in part, due to the structural constraints of the talk show itself. A traditional talk show format usually consists of a brief introduction to the show's topic of the day, followed by the introduction of guests and their stories, a short discussion with the host and audience and concluding with the expert giving advice to the guests. There are roughly five-minute segments allotted to each guest where they are supposed to tell their story. This very structure of the talk show works to individualize these social problems. The "tooc" show does not have time to explore the girls' life-stories. The talk show is a here-and-now format, decidedly situated in the present with the goal of the future. However, the inevitable failure of this approach to make a critique (or to radically alter the lives of the girls and women on these programs) is situated in the erasure of the past. In talk shows, there is a need to stay on task because of the capitalist nature of time with regards to television. The genre offers a made-to-order format allowing the repetition of similar narratives and character-types (stereo-types)—the faces and names are different, but the roles are the same. In this way, the girls' life-stories aren't mentioned because they complicate and interfere with the pre-existing format. One can only fit so much information into a five-minute segment. Similarly, the implications of race and class for these life-stories remain unaccounted for.

SELF-CONTROL

Related to individualism is this notion of self-control. Perhaps the best way to understand self-control is as a gendered norm. Self-control is an expectation as to how to behave with a definite sense of appropriate and inappropriate behavior. Deborah Tolman makes this notion clear in her essay "Adolescent Girls' Sexuality: Debunking the Myth of the Urban Girl." Tolman, citing Carol Gilligan's work, writes, "some girls consciously keep their true thoughts and feelings protected, out of relationships where saying what they think and feel can be dangerous."[57] Then, in terms of sexuality and sexual behavior, women are responsible for policing their own sexuality as well as the sexuality of men. So, if a woman gets pregnant, it is her own fault because she did not stop the man. While we all ought to know better, a *Ricki Lake* expert recently said, "Women allow themselves to be taken advantage of."

Self-control can take severe forms. As Davis and Statz point out: "In response to a husband's complaints. . .or a woman's own dissatisfaction with her lack of meekness, doctors would perform clitoridectomy, nonmedically indicated hysterectomy, and other gynecological surgeries and manipulations."[58] Bartky, in her book *Feminism and Domination*, argues, "This self surveillance is a form of obedience to patriarchy. It is also the reflection in a woman's consciousness of the fact that *she* is under surveillance in ways that *he* is not."[59] In these instances of self-control, a woman's perceived failure to meet the expected demands of society results in her policing her own body on a very physical level in order to conform.

In terms of the teen-out-of-control show, we see self-control working on two levels. The first level is the girl herself. Chances are the girls who are on the talk shows do not have the self-control expected of them. In this way, self-control is a morality issue. The second level is related to this expectation of self-control. This is what is implicitly, and sometimes not so implicitly, implied in talk show discourse. "Teen-out-of-control" guests are on the show because they have been unable or unwilling to police themselves. The quote above from the talk show expert implies that women are the controllers of not only their own sexuality and behavior, but that of men. This has become a taken-for-granted assumption with dramatic ramifications for women's views of themselves and their experiences.

BODILY CONTROL

The above discussion on individualism and professional knowledges and self-control serves to illustrate that these forms of punishment/reform are not arbitrary, but rather rooted in institutionalized discourses. When female subjects do not conform to gender norms, systems of discourse have risen up to treat and reform—either through make-overs or boot camp—and reconcile the woman/girl back within normative bounds. Foucault suggests, "in our societies, systems of punishment are to be situated in a certain 'political economy' of the body. . .it is always the body that is at issue—the body and its forces, their utility and their docility, their distribution and their submission."[60] Throughout time and across cultures women's bodies have been sites where domination and control have been enforced. Perhaps one of the most powerful explorations of bodily control comes from Judith Allen's "Men, Crime and Criminology: Recasting the Questions."[61] Allen argues for a return to examinations of the "sexed-body," recognizing it as a site of coercion and control. Allen makes an important argument. The research clearly shows that there is more to the social control of women than notions of masculine and feminine. This social control is being carried out on women's bodies, not men's (as much). The sexed body needs to be taken into consideration, especially in this research, especially that as recent as 1980 some jurisdictions submitted juvenile girls, regardless of arrest charge, to undergo examinations for venereal disease, pregnancy, or drug use.[62]

James Messerschmidt offers an historical context of bodily control in his essay "Feminism, criminology and the rise of the female sex 'delinquent', 1880–1930."[63] He begins by saying that girls are more likely to be questioned by the courts about their sexual activity than are boys. Furthermore, girls' offenses are more likely to be viewed "as an aspect of sexual promiscuity, and more likely to lose their liberty for activities which would not be against the law if committed by an adult."[64] What is interesting in Messerschmidt's essay are the examples of the ways in which bodies are regulated. He argues that the current roots of the sexualization of female deviance can be found in the movement against the legalization of prostitution. This would have required prostitutes to undergo vaginal examinations and licensing. The resulting defeat of the legalization saved the prostitutes from this sort of bodily control (no similar control was offered for men). However, the goal of the movement was to increase morality—a whole system was set up on an ideology of conservative Darwinism. In this way, "only the *moral* could attain progress, it was the *immoral* who had to be controlled."[65] Refocusing after their defeat, the reform movement made it their new goal to protect the morals of young women. Those who were sexual, including masturbation, were now seen as immoral. As a result, girls were frequently subject to vaginal examinations by the courts to prove their chastity.

It is interesting that on talk shows, chastity is no longer an issue. The women and girls aren't there to prove their chastity. The focus is on sexuality issues of pregnancy, drug use, disease, and paternity (although this is a paternity with a decided focus on exposing the mother's sexuality). Much of the literature surveyed regarding women in the justice/welfare system mention forms of bodily control—be it self-imposed or imposed from the outside. Generally, institutionalization was the main focus, but there were other forms of control alluded to but not explored. One of the most striking of these was the at home abuse many female offenders endured. Certainly, that is a form of control, and one that is apparently unofficially sanctioned by the state as indicated by their unwillingness to prosecute abusers. Instead the state tends to institutionalize the victim. This in turn supports the individualization of the girl's problems and works to erase her life-story.

REPENTANCE THEMES/RESOLUTION

In talk shows, the "out-of-control" girls are sent to boot camp or prison. When they return, they come on stage and hug the host, their mother, and the boot camp authority figure. We can see this resolution and repentance alluded to in the detailed show synopsis from *Maury* listed above. This plays out what we call repentance themes. Once the punishment/reform has been carried out, one needs to assess the effectiveness of it. Lorraine Gelsthorpe's article "Towards a Sceptical Look at Sexism" has been most useful in understanding the repentance themes in talk shows.[66] Gelsthorpe examines the use of sexism as a determining ideology in the criminal justice system by conducting ethnographic research with a British Juvenile Liaison Office. She found that girls are more often given cautions, instead of prosecution, if they appear sorry—i.e., cry. Furthermore, they are more likely to get cautions if their parents seem concerned and involved (the same is true of boys).

This resonates with the "tooc" shows, especially on *Maury*. After their prison/boot camp experience, the girls come out and give Povich a hug thanking him for his help. Some appear more grateful than others, who appear forced into this affectionate display. Regardless, the action itself is meant to symbolize a realignment with, or at least a recognition of, authority—in much the same way that crying to the police officer indicates the same realignment and recognition. Even so, one has to wonder who the intended beneficiary of the talk show reconciliation actually is. Is it just for public appearance? Is it to legitimate the use of boot camp or prison as a means of punishment/control/deterrence?

This points to how girls are more often the subject of talk shows, not only because of the history of public social control, but also of the likelihood that there will be a reconciliation with power structures and a realignment with authority. For instance, on the "tooc" show, the girls giving hugs and thanking the host and the drill sergeants for their discipline indicates that the girls recognize authority and are submitting to it. Furthermore, this repentance also reinforces the erasure of the life-story. Because the experience itself temporarily removes the girl from the environment that created her, she will return to an unaltered home life. Without altering the girl's permanent environment, the impetus of change is put on the girl to accept her life-situation, which often includes abuse and neglect. While there will be mention during the talk shows of parental drug addiction, violence, and sexual abuse, these issues are essentially confined to the home and are rarely continued or taken account of in the talk show narrative and resolution. However, because the talk show rarely explores this life-story, the girls' repentance in the final segment of the show works to legitimate the boot camp solution and bodily control. This type of punishment comes to be seen as natural and effective. The disciplines and discourses of professional knowledges and the talk show form have worked simultaneously to legitimate bodily control.

WHERE ARE THE BOYS? THE CASE OF THE TEEN-OUT-OF-CONTROL PROGRAM

We are revisiting this category to elaborate the position of men and masculinity in the teen-out-of-control program. We argue that men have a position of presence and absence in this discourse. First, their absence is manufactured by the very construction of the discourse. The teen-out-of-control is female and defined by her sexual behavior. And while a rare episode featuring a violent teen boy does air, it is an exception and can be categorized under hypermasculinity. This again refers back to Judith Allen's work on men and masculinity. One needs to understand the implications of the presence and absence of a male body in terms of authority, power, and control. We have problems conceiving of a boy-out-of-control as a teen-out-of-control, because the primary identifier of "tooc" is sexual promiscuity. When it is all said and done, we do not have the same standards of morality for boys' and girls' sexual behavior. Deborah Tolman also acknowledges that boys/men do not bear the responsibility and all-too-real consequences of morality saying the boys "are never considered promiscuous and cannot get pregnant."[67] This speaks strongly to the absence of men as subjects of this discourse. By exploring this absence, the focus on women becomes front stage. The other conspicuous absence is that of the father. Rarely do fathers participate in, or solicit the help of, the tooc program. Occasionally, an audience member will ask a guest where her father is (and guess what? She usually doesn't know). Women are the object/subject of this discourse.

However, never fear, men are not totally absent in the case of the "tooc" show. They are reinserted in the form of the male host, male counselor, and/or drill sergeant—in other words, in the form and voice of authority (they take the place of the father). The male host—specifically Maury Povich and Montel Williams—offers a rational, yet sympathetic masculinity. They will attempt to control their show according to its structure and content. In the case of the "tooc," the masculine host is non-threatening and supportive, but firm. While he may send the girls off to boot camp, it is really for their own good. *Maury* especially works in this way when the girls, upon returning from boot camp, thank him for helping them to see the error of their ways.

As mentioned above, the male counselor and drill instructors straddle the line between a sensitive patriarchy and hyper-masculinity. As they are on the side of the talk show, the law, and the mothers, the drill sergeants are working in connection with punishment/reform goals of the program. So while their actions may be harsh, they are validated by the logic of "for your own good." In this way, the drill sergeant/counselor is on the side of patriarchy, both institutional and personal. However, were it not for this context, these men would fall under the hypermasculine category. First, they are always physically large, muscular men—at times they are almost exaggerations of masculine bodies. Second, their initial tactic is to yell at the girls. Of course the girls, having no respect for authority, yell back. The drill sergeants then yell louder. They instantly try to establish their position of unquestionable authority through intimidation. And, they are not afraid of bringing race into the mix. On the "Send My Wild Teen to Boot Camp" episode of Sally, two drill sergeants, both black men, come out yelling on the stage. One of them corners a girl, yelling at her just inches from her face. He asks, "Do you know who you're dealing with?" There is a slight pause and no answer. He then yells, "Devil with a black suit on." This is clearly operating on the stereotypes associating violence with black masculinity. However, not all the yelling is as threatening as this one. Drill sergeants and male counselors alike will yell motivational threats as well. For example, Duane West, a motivational speaker and a *Maury* favorite, will yell things like, "I'm going to give you some self-respect" and "You're going to learn to respect your mother." This is a very grand assumption that he can give the girls some self-respect, but it is one tied up in the power of patriarchy and reform. Once the drill sergeants have control of the girls, they confine and regulate their bodies by forcing them to do exercises, by making them ask for permission before they may move or speak, by putting them in prison cells. These are all strategies by which the girls' control over their own bodies is diminished and subjected to patriarchal reform. By the end of the program, the "reformed" girls are brought back to the television studio where they admit their failings and submit to authority. This usually involves the girls coming out with gifts for her mother, and hugs for the host and the counselor or drill sergeants.

As the above discussion illustrates, masculinity does indeed have a presence in the teen-out-of-control program. It works as a symbol of authority, power, and reform. And, it works on the bodies of the girls. These representations of masculinity are bound up with the justice/welfare systems, which provide the rationalization for these spectacles of control. Furthermore, these representations of masculinity further emphasize the powerless mother. She bears the burden of not only being deemed a powerless mother, unable to control her child, but she also becomes a victim, the target of her daughter's wrath.

SOME CONCLUDING THOUGHTS

This chapter is an exploration into how discourse and discipline work on and through the daytime television talk show. Throughout this chapter, we have revealed how and why talk shows have an emphasis on the female body, which works to reinforce and reinscribe gender norms of femininity and sexuality.

We have begun to illustrate ways in which women are presented in the talk shows, and we articulate these presentations in a contextual way. We once again agree with Foucault when he says, "we are dealing. . .with a multiplicity of discourses produced by a whole series of mechanisms operating in different institutions."[68] Tracing the paths of these discourses as they lead from the talk show to and through these different institutions has proven to be a complex and useful task. It is useful because by acknowledging the complexity of talk shows, we get a better understanding of the subtle ways vulture culture borrows from, scavenges, and reproduces discourses found in society and its institutions.

7

ADMISSIBLE IN A COURT OF LAW:

DNA, PATERNITY, AND THE TALK SHOW

What I am going to hand over to you is admissible in a court of law. Ok? This is a true DNA test. [to the father] So, to an accuracy of 99.999% this has been your child for the whole 18 month period. [to the mother] So, since on the show he has made an admission to the fact that he has a job, you can take that down to a family court and they can garnish his wage. [to the father] It really has nothing to do with whether or not someone lets you see the child or not, or whether you have difficulty in your visitation. You are a father. And, I don't care if you never see your child again for the next 18 years. I hope that the court takes money out of your check every day so we don't have to pay for your child. —Montel Williams from *Montel*'s "Paternity Questions. . .Are the Rumors True?"[1]

Paternity episodes provide us with another opportunity to consider talk shows as materialization of vulture culture. Instead of seeing these programs as outlets for "sailor-mouthed moms and their accused babydaddies,"[2] we want to suggest that talk shows can be fruitful sites for cultural critics and provide valuable tools for media literacy. They can give us insight into the everyday assumptions imbedded in notions of fidelity, out-of-control behavior, and paternity. As such, this chapter serves three main functions. First, it explores the logic of paternity in the talk show and in our culture. Second, it continues the examination of femininity by also exploring and complicating masculinity. Finally, this chapter is the culmination of the themes of this book in a concrete example. In this chapter, an exploration of paternity shows provides us with the opportunity to investigate talk shows as vulture culture by raising questions of gender (of motherhood and fatherhood), sex, science, equality, and welfare (child support) through the examination of expert knowledges and professional discourse. By denaturalizing the logic of paternity, we provide insight into the very particular ways in which vulture culture operates.

Examining the use of the DNA test is key to bringing to light questions of paternity and of maternity, of the biological and the social. We begin to see the ways in which men and women are situated in our society as fathers and mothers. We begin to see the gendered, economic implications of "so we don't have to pay for your child." As this chapter will show, exploring the development of DNA/paternity testing and its implementation in the legal system illuminates the ways in which mothers and fathers are situated under capitalism. Furthermore, we will see how the discourses of science and of the legalities of paternity work to situate the mother as fixed and the father as flexible.

Trust, Truth and Betrayal: Paternity on the Talk Show

The paternity subgenre of the talk show has become a dominant programming staple and plays out in essentially three scenarios: First, "I know you're the father"; second, "I'm not sure who's the father"; third, "This time. . .I'm sure he's the father." Initially, when paternity shows first hit the talk show scene, they seemed to have an "altruistic" motive towards the child and the mother. The goal of these programs was to hold the father accountable for, at the very least, child support and to encourage visitation. We'll call these examples of paternity testing "I know you're the father." The DNA test results would be admissible in court as proof of paternity should the mother or father choose to pursue the matter. And, as we can see in Williams' comments above, *she* is often encouraged to do so. There are significant socio-economic implications underpinning Williams' words "so we don't have to pay for your child." For example, this quote is taken from an episode of *Montel* where 16-year-old Annie wants to prove her 18-year-old ex-boyfriend Richard is the father of her baby. Annie and her mother claim that Richard has denied the baby and has said that Annie was sleeping with other men. However, Richard argues he never said those things about Annie. Furthermore, he claims he never denied the baby, but he can not visit the baby without Annie and her mother "jumping down his throat." When the paternity test results are announced, Richard is proven to be the father. And, as we will come to see throughout the course of this chapter, there are certain expectations of a father—the first of which is to "pay for your child."

In the "I know you're the father" scenarios, the host and guests usually believe the mother. The studio audience applauds the mother and "boos" the father. However, the paternity test always presents the possibility that the mother is not telling the truth. In the same *Montel* episode, a woman named Lorraine states that her ex-boyfriend Daniel is the father of her son. Prior to revealing the test results, Williams asks Lorraine, "Now, the question becomes, who is the father of this baby? And, Lorraine, you say it's Daniel. Now you were with no one else?" Lorraine replies, "I was with no one else." Williams rephrases his question and asks, "If this baby is not Daniel's, whose baby is it?" To which Lorraine responds, "It's Daniel's Montel. I know it is. I was not with anybody else." Williams looks at the test results and says, "I bet you did know that, that's why I wanted you to say it with more force—99.999%. You're the father. Step up to the plate and do what you should've done."[3] In the end, this story proves two things: First, Daniel is the father and, second, Lorraine is telling the truth.

Despite the arguably altruistic motives of paternity shows, there has been a noticeable shift in the spectacle. This shift points to the flip side of DNA testing. Increasingly, the "I know you're the father" shows are finding that despite the mothers' insistence, the men being tested are proven not to be the fathers. In a *Maury* episode on paternity, a young woman named April insists a man named Tony is her daughter's father. She angrily yells at Tony, "I know, and I'm 100% sure. I know that she's yours. You're the only one I was with."[4] However, as the results are read, her jaw drops in disbelief. The DNA test has determined that Tony is not the father. April runs off stage—confused and humiliated. When Povich chases after her, she cries, "Leave me alone. I don't want to go back out there." So, while we may at first think of the paternity test as revealing the identity of the father (as in the cases of Richard and Daniel above), it also can serve to reveal the infidelity, promiscuity, or simple trustworthiness of the mother. The spectacle no longer revolves around the scene where the mother unleashes the "I told you so" wrath. Rather, the spectacle has shifted slightly to the moment where the test calls into question her sexuality, morality, and ethics.

More recently, paternity shows are featuring mothers who bring several men to be tested because they are unsure who exactly fathered their children. This is the second way paternity is presented. We'll categorize these examples as "I'm not sure who's the father." The increased repetition of these scenarios has altered the fundamental question of the program. Instead of the question being one of paternity, of exposing the father, the question becomes one of maternal responsibility, sexuality, and fidelity—exposing the mother. This shift is obvious in a July 2002 episode of *Jenny Jones* titled "Who Is the Father of My Baby?" During the episode's opening segment there is a voice over which says "It's a test of trust. . .it's the test for paternity. . .shocking stories of truth and betrayal. . .DNA tests prove who's the father."[5] In this description of the program, we can begin to see how the use and implementation of paternity tests have cultural and social weight, especially with words like "trust," "truth," and "betrayal." While these words can apply to both mothers and fathers, in the context of this *Jenny Jones* show, we are to understand them in relation to the mother. In four of the six paternity segments featured, two men were tested. In three of those cases, the woman was in a relationship with one of the men at the time she got pregnant. "Trust," "truth," and "betrayal"—can we believe the mother? This becomes the question of the paternity test.

Finally, the third way paternity plays out on talk shows are the follow-up or update shows. We'll call these the "This time, I'm sure he's the father" shows. These shows have developed out of the first and second type of paternity scenarios where none of the men being tested are proven to be the father. In these episodes, the shows' producers bring the mothers back with new men to test. In some cases, it is one man, in other cases it can be several. These shows have directly evolved from the spectacle of the "deviant" mother. They exist only because the women have been sexual and/or untruthful. One notable example comes from *Maury*. On this update episode, Povich recaps the story of a young woman named Tiffany who was featured on a previous episode. She believed that her ex-boyfriend Branden was the father of her son. In making her case, Tiffany told how Branden was at the hospital when she had the baby, that he named the baby and signed the birth certificate. She was left confused as to why he was now questioning paternity. When Branden came out on to the stage, he explained to Povich, "I always had doubts. Always. . . But, I tried to be a bigger man, but then I'm with a tramp like her." Tiffany angrily cut Branden off, "A tramp?! A tramp?! You gonna tell me to shut up and you gonna talk so you can call me names, you out of your mind!" However, when the paternity test results were revealed, Branden was not the father.

In this update show, Tiffany has returned to *Maury* with two additional men to test. Povich asks her, "Have you been doing some soul searching to try to find out who little Anthony's father is?" Tiffany replies, "Yes I have. Since the show, I thought about it seriously and there was two other guys that [she pauses briefly] I was with during that time." When Povich reads the results, Tiffany is again humiliated. Neither of the men fathered her son. She bows her head and stares at the floor. Povich puts his arm around her and says, "If you want to find out who the father is, we will continue to help you. Ok, Tiffany? All right?" And she nods. Perhaps we'll see Tiffany again on a future episode of *Maury*.[6]

These paternity scenarios set the stage for the remainder of this chapter. Women and men as mothers and fathers have culturally prescribed identities, responsibilities, and expected behaviors. These paternity episodes point to a rupture in these idealized parental and romantic roles. They point to instances where both women and men fail as mothers and fathers, wives and husbands, girlfriends and boyfriends. The search for paternity is, in part, an attempt to locate and remedy these failures. However, our contention is that the very notion of paternity keeps these failures and solutions trapped in the paternalistic discourse which creates them. As you will see below, the logic of paternity and the responsibilities attached to it may no longer be reasonable for today's families, if they ever were. Indeed, the very question of paternity might be the wrong question.

THE FLEXIBLE FATHER AND THE IMMOBILE MOTHER

As both hard science and television spectacle, we argue that the talk show's use of the paternity test serves the interests of capitalism using the ideologies of science and the justice/welfare system. In order to understand how mothers and fathers come to be situated through DNA testing, we need to premise three overlapping assumptions that set the foundation for the remainder of the argument. First, capitalism, as a means of organizing human relationships, is flexible and will use existing ideological and discursive formations in order to exploit labor for profit. Second, patriarchy, that dreaded p-word, is one such ideological formation. Rosemary Hennessy discusses the importance of recognizing patriarchy: "patriarchy refers to the structuring of social life—labor, state, and consciousness—such that more social resources and value accrue to men as a group at the expense of women as a group."[7] She goes on to argue that patriarchy has often been used in the service of capitalism's exploitation of labor, particularly women's labor (this will be discussed further below). It is important to note the forms in and through which patriarchy operates are variable. By this definition, patriarchy is also flexible. As such, we see patriarchy as mediating between capitalism and labor. In this way, patriarchy is crucial to understanding the social positioning of the father and the mother. Third, while capitalism is flexible, the one constant, the moment of fixity is capitalism's exploitation of labor (although certainly the forms of exploitation vary.)

With these three things in mind, we argue that on the talk show, in the justice/welfare system, and in the very "science" of paternity testing, fathers are aligned with the flexibility of capitalism and patriarchy, and mothers are aligned with the fixity of exploited labor. More specifically, we are arguing that the development and subsequent uses of paternity testing presume the mother as fixed, as immobile. Scientifically speaking, her DNA is the control group; his DNA is the variable. Judicially speaking, it is assumed that the mother will retain custody of and responsibility for raising the child; the father has a variety of options ranging from child support to full custody.[8] Still, his legal obligation can be as simple as mailing a check once a month. But, it should not seem as if science and the judicial system are working independently in this enterprise. Rather, they are intimately bound up with one another. However, in order to see how these areas fit together, we have to present them separately. Each comes with its own logic and history, which in turn contributes to the others' logic and history.

Barbara Katz Rothman in her essay "Beyond Mothers and Fathers: Ideology in a Patriarchal Society," suggests this kind of work is an unraveling of sorts. She writes, "To understand it, to explain it, we need to step back and try to disentangle the contradictions. When we do, we find ourselves unweaving the strands of a fabric, understanding the pattern as we work it backwards to the underlying threads."[9] If we imagine vulture culture as our tapestry, then we can see how science and the judicial system are key strands. As DNA testing is the touchstone in this paper, we begin our unweaving with "science," specifically technologies that produce a "visual" interpretation of the body, such as Ultrasounds and MRIs. Then, we pull on the thread of the justice/welfare system in order to see what part of the pattern is linked to the uses of paternity tests in determining the rights and responsibilities of fathers and mothers. Understanding the connections between science and the legal systems then works to illuminate and situate the spectacle of paternity on the talk show.

BLINDED BY SCIENCE

"Is science a useless exercise that can only reproduce social ideology, while offering us its metaphors for nature, whereas nature, as we conceptualize it, is only a metaphor for social reality?"[10]

When the talk show uses the paternity test, it is relying on the claims of truth, of fact, and of science that are attached to this test. However, this test and the science on which it is based are ideological in their construction and implementation. Here, we will examine how scientific discourse is ideological, especially around questions of the female body and paternity. When referring to an entire discipline, such as science, one begins to wonder where to start and where to find the point of entry. Certainly, entire books have been written about the social, economic, and political influences on the development of science.[11] Drawing off this work, we briefly present the arguments challenging positivist science's claims to an objective understanding of the "natural" world. Much of Western science claims to achieve such an objective understanding by using rigorous and tested methodologies that work to eliminate "the values, vested interests, and emotions generated by their class, race, sex or unique situation."[12] Furthermore, sociologist Patricia Hill Collins suggests that the scientists' attempts to detach themselves are mirrored by their attempts to decontextualize their objects of study (this is especially clear in the quest for paternity).[13] In challenging this objectivity and decontextualization, feminist scholar and biologist, Ruth Hubbard asks some provocative questions which set the stage for critique: "What is 'true' about nature depends on who is asking, under what historical and sociopolitical circumstances, from what point of view, and to what end."[14] The implications of scientific lenses becomes clear when Hubbard relates "scientific" accounts of female intellectual capacity: "it should not surprise us that nineteenth-century biologists, who were by definition male, found scientific reasons why girls could, or should, not get the same education as boys. Some of them 'proved' that women's brains were smaller than men's, others that education damaged girls' reproductive organs so that educated women would not be able to have children."[15] In her book *Profitable Promises*, Hubbard works to articulate and then debunk gendered constructions of science. For instance, she deconstructs the reproduction myths of the active male and the passive female by challenging popular scientific assumptions of androgen and estrogen, sperm and egg, chromosomes and DNA and the X and Y-chromosomes.[16] Nevertheless, this active/passive dynamic remains the underlying assumption in reproduction. Here, it becomes clearer how patriarchy, as a "structuring of social life," has implications even for "objective" science. Furthermore, as we will see below, the logics of patriarchy and capitalism work to frame the ways in which science frames and comprehends the body.

"MY BABY LOOKS JUST LIKE HIM"

One of the founding principles of the scientific method is observation. Implicit here is a reliance on the visual, on what can be deduced by sight. In their book on visual culture, Marita Sturken and Lisa Cartwright argue, "In the rise of the natural sciences in the nineteenth century and in biomedicine today, vision is understood as a primary avenue to knowledge and sight takes precedence over the other senses as a primary tool in the analysis and ordering of living things."[17] The talk show relies on the notion of observation-as-science as part of its spectacle. Common to virtually every paternity episode is the moment when the mother compares a picture of her child to a picture of the father (projected on large, side-by-side television screens). In this way, guests and the audience can try to determine for themselves whether or not the child was fathered by the man. Often, the mother will point to features that "look just like" the father. For example, April, who was mentioned earlier in this chapter, enthusiastically compares her daughter to Tony. She argues, "My baby looks just like him. Look at her ears! Look at her face! She looks just like him. How could he deny her? How could he deny them?"[18] On another episode of *Maury*, a young woman named Heather tells Povich and the audience, "That's his father. Look at my son. Look at him. He's got his eyes, his forehead, and the dimples. He's got my son's dimples. That is his father."[19] But mothers are not the only ones making these comparisons. Fathers will also rely on these photographic comparisons to point out dissimilarities between themselves and the children. And, in one instance, ten-year-old Mike is asked if he thought the man being tested is his father. Mike replies, "Yes. . .cause he looks just like me."[20] These examples give a quick insight into how a visual common-sense "science" works on the talk show. But, this emphasis on the visual has a significant history, especially in connection to women and their bodies.

In trying to understand scientific conceptions of the body, we can begin to see the intimate connections between science and technology. Technologies allow scientists to make the body visible. Documenting the body via visual technology is a tradition that begins with photography. Since the early 1900s, certain technologies have allowed science, particularly medical science, visual access to hidden parts of the body. X-rays, MRIs, and the assorted "scan" have provided science with a previously unavailable view of the body. And yet, unlike the paternity photographs described above, these visual representations of the body also require interpretation—expert knowledge. In this way, these images do not stand alone—their scientific value and meaning are derived through "expert" interpretations. One of the best examples of these sorts of visual technologies is the sonogram. The sonogram is an imaging technology that allows doctors to "view" the fetus inside the mother's womb. The healthy sonogram image appears as a grainy, out-of-focus, black-and-white representation of the baby (see Figure 1). In the picture below you can even see the baby's fingers and toes, but usually the image is not very distinct.

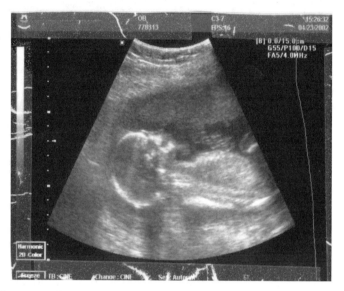

Figure 1: Sonogram (Courtesy of Mike and Kelly Cain)

Yet, the sonogram is more than just a representation of a fetus. While technology has facilitated this sort of representation, the meanings given to the fetus and the mother are socially constructed. In an essay on media uses of fetal images, E. Ann Kaplan suggests, "In terms of the science discourse, it appears that a new part of the human (in this case, specifically the female) body had come into view because technologies exist to make it visible and manipulable."[21] In the instance of the sonogram, we are not just getting a look at the fetus; we are getting a look inside the woman's body. What raises many feminist scholars' concern is that this image of the fetus is being constructed/imagined as a person separate from the woman's body. This challenges the woman's authority over her own body. Kaplan argues, "When central, the fetus renders unimportant woman's work, sex, and mother subjectivities: her body (assumed to be the home, in heterosexual marriage) is now to be in service of the fetus."[22] This link between the mother's body and the child continues to define and reinforce the socially prescribed relationships between mother and child.[23]

The image and the imagining of the fetus as person subordinates the interests and knowledge of the mother. Sturken and Cartwright suggest, "an ultrasound image taken by a doctor will be perceived as more reliable than a woman's description of her bodily sensations of pregnancy."[24] So, there seems to be a tension here. Once a mother,[25] the woman is expected to provide the care for and nurturance of the child throughout the child's lifetime; however, the ways in which she cares for the child are subject to various institutional and authoritative knowledges. This care and nurturance is generally understood as a biological instinct, a given, rather than a socially constructed behavior. And, as a biological, natural instinct, the mother's labor, which sustains and reproduces the labor force, remains unpaid and unrecognized as labor, while at the same time being constantly suspect to intervention by social and medical institutions. In this way, technologies offering a visual representation of the body can work in the interests of capitalism through the common-sense logic of patriarchy. Couched in the discourse of science, such images are "held to present accurate, self-evident proof of certain facts."[26]

"TO AN ACCURACY OF 99.999%"

The use of genetic and DNA technologies mark an interesting development in geneticists' quest to fully comprehend the human body. Whereas the imaging technologies listed above work to articulate the "hidden" body (for example, bones, brains, and babies), DNA technology works to present the previously unimaginable body. The science of DNA claims to be able to chart the body, both physically and mentally. There will be no aspect of a person's body and personality that would not be accounted for in their genes. Thus, a person is reduced to a genetic map, which will serve as a quantifiable marker of a person. Needless to say, the implications of such a science are both amazing and frightening. Debates have been raging over issues of cloning and eugenics—disease prediction/ control, racial marking, and genetic selection. Concerns have been raised over the potential use of genetic information by employers, insurance companies, and the state. While these debates are fascinating, they are beyond the scope of this chapter.[27]

What is important to understand about genetics is that despite its universalizing tendency, the avenues being explored and the genes being examined point to social, political, and economic imperatives. Scientists looking for "criminal" genes, poverty genes, forensic evidence or genetic explanations for birth disorders are usually doing state-sponsored and/or corporate-sponsored research.[28] What to look for, where to look for it, and who funds it are not scientific questions. They are political and economic questions, with political, economic, and social ramifications.

In all the debates we've reviewed regarding DNA and genetics, none of them were overtly concerned with determining paternity. Perhaps, it is because, unlike hunting for the genetic marker of diseases, paternity is really just a process of elimination followed by probability calculations. Or, perhaps, the underlying assumptions guiding the search for paternity are not being questioned; indeed, current research on paternity issues seem to revolve around making it more accessible and profitable.[29] However, we argue that looking at paternity, searching for the father, ignores the material, lived social conditions of single mothers and their children. While the test may prove who the father is, the tests do not make him pay child support or schedule visitations. However, the underlying logic of the tests—the search for the paternal truth—is based on the patriarchal assumption that the man will be responsible for his child and his or her caregiver. Over and over, we witness this assumption played out on the paternity show. Whether it is Montel Williams telling Daniel to "step up to the plate," Maury Povich telling a 16-year-old to "take responsibility" for his baby, or the fathers promising to start paying child support, most talk show participants acknowledge the father's financial obligation to the mother and child. It is based on another assumption, and not necessarily an incorrect one, that the mother cannot financially support her child without additional support, either from the father or from the state.[30] Situating the mother in this way once again positions her as fixed, dependent, and immobile. Alternatively, we could imagine, at the very least, finding ways to enable single mothers to raise their children without relying on the support of the father. We could offer options from safe, affordable childcare to quality, universal healthcare. We could benefit from community support networks and employers sensitive to the needs of single parents. These things are possible, but we would first have to step outside the very logic of paternity and its connection to the maintenance of the state.

The actual form of the paternity test itself also reinforces the fixity of the mother and the flexibility of the father (see Figure 2). The child is the central part of the test. The mother is the control group. The "alleged" father is the variable. The child gets one half of her DNA from each parent. So, in the lab, once the child's DNA is mapped out, it is compared to the mother's DNA. All of the common markers are eliminated from the total list of markers. The remainders are then compared with the "alleged" father's DNA. If the markers match up, the man is no longer "alleged." Of course, the process is a bit more complicated. There are samples to collect, centrifuges to run, and probabilities to calculate.

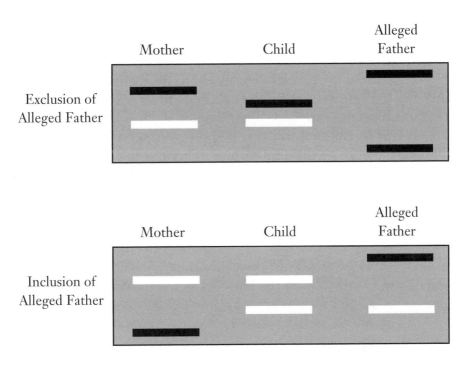

Figure 2: Paternity Test (Created by Wanda Knight-Griffin)

Nevertheless, in order for this test to work, the mother has to be the standard. Is it a coincidence that her position in this test as the control for her child mirrors her social role as mother? The father is always "alleged," he is not the father unless his markers match up. His rights and responsibilities as a father do not take effect until he is proven to be the father.[31] When thinking of the structure of this test, it is interesting to flip the terms and the logic. What might it mean to be an "alleged" mother? If paternity is determined through the DNA test, when and how is maternity determined? These questions seem ridiculous and obvious. But, they are important to ask. They point to "alleged" biological imperatives that work to situate the mother as caregiver. Sociologist and fathers' rights proponent David Popenoe uses this sort of biology to rationalize social differences between men and women and to encourage marriage: "The source of this sex-role difference can be plainly stated. Men are not biologically as attuned to being committed fathers as women are to being committed mothers."[32] Popenoe goes on to argue that women need to encourage men into marriage in order to tame them and train them to be "committed" fathers. Such logic only works to re-inscribe the mother's position as caretaker, not only of children, but also of fathers.

While there are heated debates about motherhood's relation to imaging reproductive and particular genetic technologies, paternity tests remain unquestioned. They remain a form of science where the laws of the natural world still apply or at least go uncontested. The very design of the test betrays its patriarchal roots. Paternity testing has been developed and implemented with the assumption of the mother not only as caregiver, but also as accessible. After all, the standard form of the test would not work without her DNA.[33] The relationship between mother and child is concrete and material. By its own design, the paternity test positions the father outside of the mother and child. The father's relationship to the child is always "alleged" and contingent on the leftovers of the mother-child pair.

"So we don't have to pay for your child"

The paternity test is a product of patriarchal knowledge. As such, it carries its patriarchal baggage with it as it offers itself as not only an empirical scientific document, but as a valid legal document. The tests gain legal significance when they are used in paternity and child support hearings. As such, paternity testing has become a big business. And, much of this business is coming from the state. Not surprisingly there is an intimate link between the justice/welfare system and DNA-testing facilities. For example, a lab in Denver performed 90 paternity tests in 1990. In 2000, the lab did 1,020 tests. Part of this dramatic increase comes from the lab's contract with the City and County of Denver's Department of Social Services.[34] As one journalist suggests, "Because every effort is made to find a father in the thousands of child support cases it handles each year, Denver's business alone would keep the testing center hopping."[35] However, the state is not the only client of these labs. One lab located in Fairfax, Virginia, markets its services as admissible in court and conforming to accreditation guidelines for collection and testing. They offer worldwide collection sites (including prisons) and expert witness and deposition services for an additional fee. And, their Web site features a paternity calculator "to determine the most probable date of conception."[36] Another lab located in Fairfield, Ohio, lists which popular media outlets used their services—among them are *Maury Povich*, *Ricki Lake*, *Montel Williams* and *Sally Jesse Raphael*.[37]

Parents can seek to establish paternity for a number of reasons. However as states begin to mandate the establishment of paternity, we can foresee increased use of DNA testing in child support cases. In the eyes of the state, establishing paternity legally means having a father's name on the birth certificate. In signing his name, the man claims the child, and is liable for the child's welfare. Some child support analysts suggest that mothers be rewarded with assured child support benefits if they establish paternity for the child at birth.[38] However, the state's interest in establishing paternity is not merely out of concern for the child, but also to keep from using state money to support the child.

Essentially, there are two forms of child support. The first is public child support. This would come directly from the state in the form of welfare and other benefits and is what Montel Williams refers to when he says "so we don't have to pay for your child." The second is private child support. This is paid by the "non-resident parent," usually the father,[39] and is managed by the courts, possibly resulting in garnishment of the father's wages. When paternity of the child is established, the state hopes to insure that the father will support his child at least financially. And, through the development of child support enforcement programs, there has been an increased push to determine paternity. However, many critics of child enforcement programs argue that these programs are "primarily being used as a way to get poor fathers to pay back the welfare system."[40] These critics argue that the money is not making it to the children but rather is being recycled back into the system.

The paternity test serves two functions in the legal system—to claim, enforce, or deny paternal rights and responsibilities, and to maintain the fixed position of the mother as dependent. Briefly mentioned above, in determining paternity, one is also establishing a social relationship between father and child. This gives the father certain legal and moral responsibilities, but no one can seem to agree as to what those specific responsibilities are other than the financial support of/for his child. Popenoe argues, "fathers should be taught that fathering is more than merely providing food, clothing, and shelter to children and letting mothers take care of the rest."[41] Fathers need to be involved in their child's social and psychological development as well. However, Popenoe's use of a gendered biology already sets the terms for the father's involvement.[42] In his essay, he argues that fathers and mothers are biologically tuned to rearing children in particular, "complementary" (read opposite) ways. So, the father would not provide the same nurturance as the mother. As such, Popenoe's construction of responsible fatherhood fits with the rhetoric of the father's rights movement. Critical legal scholar Drucilla Cornell makes clear the assumptions implied by Popenoe: "Simply put, the fathers' movement does not want men to parent— they want them to *father*, and they have very specific ideas about what fathering entails. First and foremost, it means the persistent reinforcement of the rigid gender divide in the family so that men can rest assured that they will not be 'femmed' by their acceptance of the role of the good family man."[43] At the same time, there have also been popular discourses opposing these biological conceptions of fathers, which work to construct a more egalitarian parenting relationship. However, with conflicting discourses of fatherhood, with biological discourses about virile masculinity underpinning much of our understanding about fathers, fatherhood, and their role in the patriarchal family, the individual man is left without much guide or direction. In this way, the flexibility of the father is both liberating and debilitating. Liberating because, unlike the mother whose relationship to her child is fixed, the father has a certain degree of flexibility in determining the parameters of his relationship with his child. Indeed, the father has a greater flexibility to decide whether he wants a relationship at all with his child. Furthermore, this flexibility is also related to greater economic wealth, because they do not have the day-to-day expenses and time demands of children. But, this flexibility is also debilitating. Without an agreed upon role, fathers (and mothers) remain uncertain as to what the responsibilities of fatherhood entail.

Mothers in the justice/welfare system do not fair any better. Cornell points out that "[d]espite many of the nations reforms of women's standing, family law has yet to fully separate women's legal identity from the enforcement of duties in the family."[44] In the eyes of the state, the very point of determining paternity is to establish support for the child and mother, the assumption being that the mother, without financial support from the father, will be reliant on the state. The woman is seen primarily in her role as mother, as nurturer, as the producer of good citizens, a role already existing in popular discourse and reaffirmed by the structure and use of the paternity test. As such it is in the interest of the state, through the justice/welfare system to assure that the child and mother receive the resources necessary to the development of good citizens. Furthermore, the positioning of the mother as dependent is based on a liberal definition of citizen which claims good citizens do not take more than they give to the state. However, mothers who can not contribute to the state, who instead are reliant on the state for the everyday support of their families, are considered dependent on the state.[45] By continually positioning the woman as mother, by both encouraging and demanding that she fulfill that role, she is positioned as to always be dependent. Her mothering labor as unpaid labor takes precedent over her paid labor. Until we work to change women's access to family-wage, flexible employment, childcare and healthcare OR until we decide to compensate women for their "mother" labor, many women as mothers will have a struggle to balance the financial, emotional, and social needs of herself and her children. She will have to petition the state to demand support from the father, or she will have to rely on the state for assistance.

Just as paternity testing sets the mother as the control and the father as the variable, in paternity cases, the justice/welfare system assumes the fixity of the mother and the flexibility of the father. These systems mirror each other, reinforcing each other through patriarchal common sense rooted in biological assumptions about women and men. In the instance of paternity, science and the justice/welfare system prove to be what Rosemary Hennessy calls "capitalist patriarchal structures." These structures "help to secure an exploitative system of social differences by way of ideologies of gender that naturalize and reproduce the asymmetrical social divisions that help to sustain, manage, and maximize the appropriation of surplus labor through a variety of complex arrangements."[46] We can see how science sees its objects through patriarchal lenses. We can also see how the justice/welfare system also relies on and reinforces these lenses. The paternity test becomes implicated in both systems through its development and implementations. These systems become reciprocal and mutually constituting. The logic becomes hegemonic, and hence, is unquestioned as it is taken up by popular culture. This interplay is vulture culture at work. Hence, talk shows as vulture culture present us with more than paternity tests as innocent and interesting additions to the talk show form. These shows reconstitute an appeal to paternity tests which is couched in a long patriarchal tradition.

"Paternity Questions — Are the Rumors True?"[47]

Having provided an examination of the institutional discourses and structures underpinning the talk show, we now want to return to the shows themselves to address three common talk show moments which further illuminate the fixity of the mother and the flexibility of the father: First, the verbal accosting of the mother by the man/men being tested; second, the mothers articulating the responsibilities of the fathers; and third, the responsibility placed on the mother to encourage the relationship with the father. While these moments are related to and influenced by each other, they represent different moments in the narrative of paternity. An analysis of the paternity talk show highlights the questions of responsibility—both the father's responsibility to the child and the mother's responsibility to name the father of her child, and questions of welfare/support.

The first moment is the flat-out verbal abuse of the mother by the man being tested. This usually occurs in the form of a pre-taped clip played after the woman has made her case to the host and before the man comes on stage. And, sometimes it is played again before the test results are read. One particularly offensive instance occurs on an episode of *Maury* and will serve as a good introduction here. On this episode, Terry, the father, has an especially violent, aggressive and demeaning diatribe to unleash on Connie, his ex-girlfriend and the baby's mother. Terry says:

> Connie, I'm 150% sure I'm not the father of your kid Niklos. Connie, our relationship was out of pure convenience. I needed a woman to be my whore and you were my whore. You're a lyin' piece of "BLEEP!" You know that kid ain't mine. And you better stop harassing me. Connie, you're a money hungry bitch who used me for everything I got. I know you slept with six other men and a girl. What's up with that? You're nothing but a ho. The kid has red hair. He's paler than Casper the Ghost. I'm not the father! You're nothing but a manipulating bitch.[48]

Another instance occurring on *Maury* is connected to the story featuring Heather mentioned above. In this story, the father with dimples Chuck says, "I'm 120% sure that I'm not the father of Heather's baby. She thinks she can get my money, my change and my loot. Heather, youze a bitch and an undercover freak"[49] And still in other examples men use accusations such as "scandalous troll," "low-down, lying, conniving bitch," "crazy, money-hungry bitch," and "you're a compulsive liar you stupid bitch."

The repeated use of the term "bitch" and "whore" in these attacks has become commonplace, part of the talk show discourse. But, we have to wonder, what makes it acceptable to treat anyone this way? Why are the men on these programs given a forum to debase the women who are often the mothers of their children (five of the six speakers quoted here are the father)? We can speculate that it may have originally been done to emphasize the moment of the "I told you so" wrath when the mother proves she was right. However, these open and encouraged tirades once again position mothers socially, politically, and economically. Regardless of the type of paternity show, the verbal assault on the mother is the first step positioning the mother as "bitch," as untrustworthy, as a bad mother, and, perhaps most importantly, as powerless. The woman's very presence on the show calls her behavior into question. This speaks to the powerlessness of women as mothers who must subject themselves to this narrative in order to legally identify the father of their children in hopes of getting support.

This takes us to the second talk show moment where the mother is asked to articulate her expectations of the father. Most often the mother wants three things from the father: financial support, emotional support, and a "father" for her child. In the story featuring Connie and Terry, Connie tells Povich that Terry only began denying paternity when Connie started asking for child support. Terry tells Povich, "That ain't my damn kid and I'm not payin' for something that ain't mine." To which a visibly upset Connie replies, "You know what? You're paying for him because you signed the paperwork. He's your kid." Later in the show Povich asks Connie, "What do you want from him if he is the father?" Connie replies, "Either A. be a full-time father as much as he can or B. walk away and don't ever contact us again."[50] On the same episode, a mother named Tabitha says that she wants a father present for her daughter: "I never thought I was going to be a single mom. . .He's missed her Christmases, her birthdays, her first words, which were daddy. . .After the DNA test proves she is his, I want him to stop denying her."[51] But, it is April who says one of the more frequently used statements regarding a father's responsibility: "I've tried to make him be a father. I didn't make her on my own. I'm not gonna raise her on my own. We made her together."[52]

In these instances, the talk show spotlights the flexibility and confusion of fatherhood. Who is a father? How can one "make" a man be a father? What is the father's role in raising a child? And what are the implications of Tabitha's daughter's first word being "daddy"? Certainly, in these cases, being a father means paying for the child. But it also means something else. Connie wants Terry to be a full-time father, while Tabitha and April both express the need for a partner in raising their children. They want the men to be active in their child's life as well as lessen the parent role on themselves. And, Tabitha suggests that her child expresses a need for her father. But here, the father gets the flexibility of deciding what his relationship will be. In some paternity episodes, the father, once determined by the test, wants custody, as in the case of Lorraine and Daniel quoted above. In other episodes, the fathers agree to pay child support but do not want a relationship with the mother or child. However, most often, fathers choose a middle ground. And yet, in all these talk show cases, the only decisive action the mother can take is to petition for child support; her relationship to the child is already determined. In this way, the woman remains fixed as mother, as dependent, as needing support. Again, the very logic of paternity necessarily positions the mother as fixed with limited choices.

The third moment comes after the test proves the man is the father. Often, the man experiences the woman's in-your-face, "I told you so wrath." At which point, the woman may call the man names, yell, point and dance, sit in the chair and smile smugly, or cry.[53] And, yet, once order resumes on the stage, the hosts inevitably undermine this moment of victory. For instance, on one episode of *Montel*, Williams has an especially difficult time with one mother. She is young, disrespectful, and under attack. At the end of the episode Williams says, "I hope that there are teenage girls watching the screen right now. I really do. I hope you're watching and paying very close attention because you can be as much a reason why your baby does not have a father as the man who's not taking responsibility."[54] Povich expresses a similar sentiment on *Maury* where he says to one woman, "If you want him to be a part of this child's life, do you think he's going to do that when you get into his face?"[55] And so it seems that women have to do more than prove paternity in order for their child to have a father. They have to be nice, amiable, appeasing. This resonates with our earlier section on self-control in the teen-out-of-control chapter. The woman is responsible, then, for her sexuality as well as the man's, and for encouraging the father-child relationship. In this way, the woman is continually asked to suppress her feelings and desires in both her romantic relationships and parental relationships (in this case, to do "what's best" for the child).[56]

An examination of the talk show text contextualizes, and renders material, struggles over paternity and patriarchy. For instance, it illuminates how fathers are able to free themselves from their children in ways that are rarely available to mothers. And finally, it shows how most of the mothers are looking for more than a paycheck. The mothers on talk shows are looking for a father who will not only be nurturing and supportive of the child, but will help her in raising the child. Each of these talk show moments reconstitutes the already existing positions of mothers and fathers. Indeed, the paternity shows are an expression of vulture culture, scavenging off of the legal, scientific, and economic logic of patriarchy. The moments discussed above both inform and are informed by the ways the gendered body, in this instance, woman-as-mothers and men-as-fathers, are constructed by science, the justice/welfare system, and popular culture.

MONEY-HUNGRY AND MANIPULATIVE, LYING AND CONNIVING: THE WELFARE QUEEN

This chapter has centered on the gendered-body, as we believe paternity first necessitates and creates such a body. Furthermore, there has been some discussion above regarding class in terms of the economic positioning of mothers versus fathers. What remains to be addressed here is the raced body. As mentioned in chapter 6, the raced body is the disavowed body of the talk show. Even as we construct paternity primarily as a function of patriarchy and the female body, the effects of paternity's hegemony have dramatic implications for the raced body. For instance, black female bodies have historically been "portrayed by politicians, sociologists, and others in the postwar period as unrestrained, wanton breeders, on the one hand, or as calculating breeders for profit on the other."[57] One need only recall the debates about the "welfare queen" to get a flavor for the political rhetoric attached to this image. The social and political distain attached to the raced female body sneaks into the talk show discourse through the very specter of the "welfare queen," (money-hungry and manipulative, lying and conniving). Merging this historicized account of the raced body with the above accounts of the gendered body, especially mother-as-citizen, we begin to see the discourses and processes by which these women become "bitches," "whores," and "scandalous trolls." The welfare queen's relationship to the state mirrors these mothers' relationships with the fathers.

In the liberal multicultural discourse of the talk show, the playing field is leveled such that all the women are suspect, subject to the same scrutiny as the raced, female body. In our previous chapter, we briefly discuss how illegitimacy was once seen as a fixable psychological problem for white women and a sign of inherent moral inferiority in black women.[58] These pop culture representations no longer treat white women and black women differently. Instead, this has become a debate about the classed female body who must subject herself to the father, the state and the talk show in order to get assistance in raising her child. White, Hispanic, black and mixed-race women all live with the legacy of the welfare queen.

THE TALK SHOW AND THE LOGIC OF PATERNITY

A survey of three weeks of *Jenny Jones*, *Ricki Lake* and *Maury* from the Fall 2001 season reveals that nearly a quarter of all the programs aired were paternity shows. The paternity show became a staple of the talk show form, and this fascination with paternity is threefold. First, as discussed above, it has a high spectacle value. The narrative revolves around a moment of revelation—is the mother telling the truth? Is he the only one she's been with? Are these the only possibilities? Did she bring in the right man this time? Is the mother a trustworthy, honorable woman who has been wronged or is she a "manipulating bitch"?

Second, the fascination and spectacle are bolstered by science and technology. As the cost of paternity testing technology decreases, these tests are becoming an inexpensive way of determining who has to pay for the child. But these are also tests of truth, trust, and betrayal. The use of paternity testing on the talk show mixes science with society. As mentioned above, sociobiological understandings of gender inform the paternity test. These tests carry more than just proof of paternity—they define the women and men as particular kinds of mothers and fathers. Indeed, DNA technologies may have a crucial role in determining social policies of the future. With this in mind, Ruth Hubbard offers some insight and a warning about the possibilities for sociobiology and genetics:

> Of course, it is far easier to blame people's genes than to abolish the sources of victimization. And it also generates profits for the scientists and biotechnology companies that are developing and marketing the tests. But if we really want to improve public welfare and health, we must

come out of the genetic house of mirrors, look at the realities that stare us in the face, and do something about them.[59]

Third, underlying each of these talk shows are questions of fatherhood, the logic of paternity, and traditions of patriarchy. These questions, logics, and traditions manifest themselves in the family and in the roles assigned to mothers and to fathers. But why the current interest in paternity? Why the search for the father? Aside from the technology and the spectacle, the logic of paternity works to reinstate the role of the father. In 2000, *The CQ Researcher* put forth one of the more interesting arguments circulating about the current resurgence of fatherhood initiatives and its underlying logic of paternity. In this article, David Blankenhorn, author of *Fatherless America*, argues that fatherlessness is a social issue of concern to all races and suggests that fatherlessness is a sign of social decay.[60]

In 1965, a report issued by former senator and then assistant secretary of state Daniel Moynihan claimed that "[a]t the heart of the deterioration of the fabric of Negro society is the deterioration of the Negro family."[61] At a time when the nuclear family structure was the white norm (more than 90% of white families were nuclear families), Moynihan and others at the Department of Labor noticed an increasing trend in the African-American community. They found that nearly a quarter of all marriages ended in divorce and a quarter of all births were to unwed mothers. Acknowledging the centuries-long systematic oppression of African-Americans and the cycle of poverty put into place by this oppression, the report calls for a strengthening of the family. Such faith was put in the nuclear family structure that it was deemed the site for government intervention. If only the family would stay in tact, the cycle of racism, of job discrimination, and of welfare dependence would eventually disappear. The presence of the father became the solution to national reform. However, this report was viewed harshly by civil rights activists and other social leaders who argued the emphasis should be on programs to reduce poverty and racism.[62]

Blankenhorn argues that the resurgence of fatherhood initiatives has to do with the fact that in the 1990s the rate of divorce among white families and the rate of births to single white mothers reached the same rates as black families and single black mothers in 1965.

Using these statistics, sociologists, minority leaders, and politicians have put the issue of fatherhood back in the spotlight. And, once again faced with the overwhelming task of reform, families and fatherhood are offered as sites of intervention. As in the talk show, the issue of fatherlessness is no longer confined to the African-American community. It has evolved into a "colorless," although certainly not classless, social issue. Positioning the struggles of single mothers under the frame of fatherlessness reconstructs the terrain of these struggles. The solution moves away from giving aid and support to the mother to resurrecting and reintroducing the father. And now, almost 40 years after the Moynihan report, the state has the technologies to do just that. Paternity testing has become the tool of resurrection and reintroduction. It resurrects the father through science and reintroduces him through the justice/welfare system. In this way, poverty, racism, and inequality become the burden of individual families, not society at large. The logic of paternity all makes sense when played out on the talk show stage where individual stories stand in for larger cultural concerns.

Understanding how paternity is constructed as a social concept, a science, a legal document and entertainment allows us to see yet other ways in which vulture culture works in and through the talk show. Vulture culture feeds on institutional knowledges, wielding them for legitimacy, credibility, and authority. It relies on the seemingly implicit and unquestioned assumptions of patriarchy and paternity as discussed in this chapter. A critical analysis of talk shows as an expression of vulture culture allows us to think through talk show narratives and question their origins, foundations, as well as rationale. Focusing on paternity issues enables us to clarify the close, though not always visible, relationship between science, law, and entertainment. In turn, this facilitates the understanding of the connections between information, entertainment, and social, political, as well as economic institutions. This understanding is important because how paternity is defined in and by these institutions has serious implication for welfare policy, "family values," and the everyday lives of mothers, fathers, and children.

8

CONCLUSION

V ulture culture. The chapters of this book illustrate the multiple ways vulture culture operates through daytime television talk shows. Based on a political and economic system of ownership, regulation, and production, talk shows constantly scavenge off of knowledges and information from different media sources, programs as well as institutions and people's everyday experiences.

There is fusion of common sense with expertise on the daily stage of the shows. There is heavy reliance on stardom packaged as wisdom and applied to concerns ranging from the personal to the professional. There is a glorification of bodies but an unwitting attempt to control and police these bodies. There is the commodification of the intertext that creates spectacular marketing vehicles. There is the convenient marriage of science, law, and family drama to solve the "mystery" of paternity. All these are reproduced in the form of infotainment. Vulture culture.

Our goal in writing *Vulture Culture* is to uncover the intricate processes by which power and control are maintained through the daily shows. Whether the shows are centered on the dilemmas of a mother with an out-of-control teen, a husband with an unfaithful wife, a son with a drug-addicted mother or a daughter with a physically abusive boyfriend, the voices of the show host, expert, and even studio audience are often the voices of unquestioned social norms.

With rare exceptions, these are the voices that write the script on what can be discussed on the shows, by whom and with what resolutions. They construct the rationale of the programs in such a way that alternative ways of thinking and acting are automatically excluded if not thought of as an aberration. Normalizing behavior, calling for conformity is the message that is repeated, over and again, despite the diversity and differences among the shows. This repetition and reliance on the script is a driving force of vulture culture. We have demonstrated throughout chapters in this book that talk shows rely on the narrative of the individual isolated from its everyday social reality. Isolation of topics aids in insulating them from the sociopolitical context in which they should be discussed, debated, and resolved. For instance, the "teen out of control" is the focus, the beginning, and end of the show. She is blamed and shamed; held fully accountable for the pain inflicted on her family, herself, and on society at large. However, the teen is usually out of context on the shows: the programs erase the forces of exclusion inherent in the very social structures that define her life. Broader issues of injustice and exclusion are thus avoided because the social systems that maintain them are hardly ever called into question. The individual bears the responsibility for any failures, crises, and/or alienation experienced.

But, people need to take control of their lives, the shows regularly remind us! Individual responsibility is the expected and repeated logic on the shows. However, the linkages between individual lives and the larger sociopolitical and economic systems that define the possibilities of these lives are rarely ever established. This is precisely how power and control are maintained by vulture culture and systematically reproduced through the daily repetition of these shows.

Our objective in focusing on daytime television talk shows is not to demonstrate ways in which these daily programs tease us, betray us, or even rob us of possibilities of meaningful dialogue. Rather, the aim is to situate these programs within the much wider culture that turns potential spaces of communication and dialogue into commercial stages of entertainment and spectacle. Again, our focus on these daily programs has provided us with the possibility of seeing how various institutions—entertainment, advertising, science, welfare, family—intersect with each other and reinforce each other. In their intersections, they promote and re-inscribe the norms and interests in the media we consume on a daily basis.

We encourage the readers of *Vulture Culture* to be not only attentive to daytime television talk shows but to other media products that also participate in the construction of vulture culture. Vulture culture is not limited to television talk shows. It is a culture that includes such increasingly popular media products as court room reality shows (*The People's Court, Judge Mathis, Divorce Court, Judge Judy*), star creation reality shows (*America's Next Top Model, American Idol, Star Search*) prime-time make-over shows (*Extreme Make-Over, The Swan, Queer Eye for the Straight Guy, Inked, I Want a Famous Face*), family modification shows (*Nanny 911, Trading Spouses, Supernanny, Wife Swap*), entrepreneurial reality shows (*The Apprentice, The Cut, Hell's Kitchen*), celebreality shows (*Surreal Life, Dancing with the Stars, Celebrity Fit Club, Breaking Bonaduce*) and news magazines (*Dateline, 20/20,* and *60 Minutes*).

What these programs have in common with the talk show is the spectacularization of the personal, the selling of intimate stories, family dramas and conflicts, and ordinary life experiences of failure and success. The programs reinforce and complement each other while presenting us with the forms of ritualized carnival discussed in chapter four. Most important, perhaps, is the reliance of all these shows on the knowledge of an expert, whether the expertise is granted through credentials, stardom, or life experiences.

As we begin to understand the complex ways in which vulture culture operates, we must move towards praxis. Critical media literacy[1] exemplifies such praxis, and thus extends the concerns of academic research to active participation in the struggle for democratic social change. For example, by tracing the relationships between paternity tests as scientific and legal documents, we have demonstrated the ways capitalism and patriarchy work through language and images to naturalize the roles and expectations of mothers and fathers. This knowledge provides a valuable foundation for rethinking the very logic of paternity. Once this logic is recognized as a constructed and constraining force, then we begin to see through its material implications and impacts on social policy.

So, these paternity shows offer much more than a mere entertainment episode of daily programming. They provide an opening for media researchers, cultural critics, teachers, and social workers to explore such questions as how knowledge is constructed, expertise defined, and life experiences framed by realities more complex than what the shows visibly admit.

This book, therefore, opens new avenues of inquiry in terms of both the theory and forms of vulture culture and raises further questions that could push the discussion and praxis further. Change can be negotiated through academic challenges and critical inquiry by exposing the social, political, and economic forces which work against justice and equality. However, we must be willing to make the challenge and have the adequate language and theoretical tools needed to break open the potential for critique inherent in the study of vulture culture.

NOTES

CHAPTER 1

1. Tara Burghart, *Marketing Execs Eye "Oprah" Giveaway* (Netscape Network 2004) [cited September 23, 2004]. Available from http://cnn. netscape.cnn.com/ns/news/story.jsp?flok=FF-APO-1333&idq=/ff/story/0001%2F20040914%2F2100437567.htm&sc=1333&photoid=20040902WXS104.
2. Ibid.
3. *A Dr. Phil Primetime Special: Family First.* CBS.com 2004 [cited September 23, 2004]. Available from http://www.cbs.com/specials/dr_phil/.
4. Vincent Mosco, *Political Economy of Communication* (London: Sage, 1996) p. 25.
5. As articulated by Douglas Kellner, *Media Culture: Cultural Studies , Identity, and Politics Between the Modern and the Postmodern* (New York: Routledge, 1995).
6. As articulated by Joe Kincheloe and Shirley Steinberg, *Changing Multi-culturalism* (London: Open University Press, 1997).
7. See also Henry Giroux, *Disturbing pleasures: Learning Popular Culture* (New York: Routledge, 1994); Kincheloe and Steinberg, *Changing Multiculturalism*; Joe Kincheloe, *Critical Pedagogy: A Primer* (New York: Peter Lang, 2004).

CHAPTER 2

1. Dallas Smythe, 2001.

2. Douglas Kellner, *Critical Theory, Marxism, &Modernity* (Baltimore: Johns Hopkins University Press, 1989), 182.

3. Ibid., 178.

4. Ibid., 179.

5. SeeVincent Mosco, *The Political Economy of Communication: Rethinking and Renewal* (London: SAGE, 1996); Ron Bettig and Jeanne Hall, *Big Media, Big Money: Cultural Texts and Political Economics* (Lanham, MD: Rowman & Littlefield, 2003).

6. Mosco, 1996, 172.

7. Ibid., 175.

8. See Ronald Bettig, *Copyrighting Culture: The Political Economy of Intellectual Property* (Boulder: Westview, 1996); Janet Wasko, *Hollywood in the Information Age* (Austin: University of Texas Press, 1995).

9. Bettig, 1996.

10. http://disney.go.com/home/today/index.html; see also Janet Wasko, *Understanding Disney: The Manufacture of Fantasy* (Cambridge: Polity Press, 2001).

11. Wasko, 1995.

12. "About the Show," 2004.

13. Among other properties—see Wasko, 2001.

14. See Bettig and Hall, 2003.

15. See Bettig and Hall, 2003, Wasko, 1995.

16. Leo Bogart, "What Does It All Mean?" In Nancy J. Woodhull and Robert W. Snyder (Eds.), *Media Mergers* (New Brunswick/London: Transaction Publishers, 1998), 17.

17. Christine Quail, *The Political Economy of Multiutilities.* Doctoral dissertation, University of Oregon, 2003.

18. Nicholas Garnham, *Capitalism & Communication* (Newbury Park: SAGE, 1990); Murdock and Golding 1974.

19. See Bagdikian, *The Media Monopoly.* 6th edition (Boston: Beacon Press, 2000); Wasko, 1995; Bettig, 1996; Herbert Schiller, *Culture Inc.* (New York: Oxford University Press, 1989); Murdock and Golding, 1974.

20. Arthur B. Kennickell, "A Rolling Tide: Changes in the Distribution of Wealth in the U.S., 1989-2001," Table 10 (Levy Economics Institute: November, 2003).

21. See Domhoff 2005; Bagdikian 2000.

22. See Albarran, 2002; Wasko 1995; Murdock, 1982.

23. SeeRobertMcChesney,"TheGlobalStruggleforDemocraticCommunication." *Monthly Review*, 48. 3 (1996): 1–20; Robert McChesney, *Corporate Media and the Threat to Democracy* (New York: Seven Stories Press, 1997a); Robert McChesney, "Digital Highway Robbery," *Nation*, April 21, 1997b, 22–24; Janet Wasko and Vincent Mosco, *Democratic Communication in an Information Age* (Norwood, NJ: Ablex, 1992); Nancy J. Woodhull and Robert W. Snyder, *Media Mergers* (New Brunswick/London: Transaction Publishers, 1998).

24. Kellner, 1990.
25. Ibid.
26. Lisa McLaughlin, "Chastity Criminals in the Age of Electronic Reproduction: Re-viewing Talk Television and the Public Sphere," *Journal of Communication Inquiry*, 17. 1 (1993): 41–55.
27. Ibid., 54.
28. Alessandra Stanley, "Mother Russia Meets Dr. Kinsey on TV Talk Show," *New York Times*, Nov. 14, 1997, 1.
29. Herbert Schiller, "Not Yet the Post-Imperialist Era," *Critical Studies in Mass Communication* 8 (1991): 18–19.
30. Colleen Roach, "Cultural imperialism and resistance in media theory and literary theory," *Media, Culture & Society* 19, (1997): 47–66; Schiller, 1991.
31. Wasko, 1995.
32. Schiller, 1989.

CHAPTER 3

1. Karl Marx, "Capital: Volume One," In Rober C. Tucker (ed.), *The Marx-Engels Reader*, 2nd edition. (New York: Norton, 1978), 321.
2. Ibid.
3. Vincent Mosco, *The Political Economy of Communication: Rethinking and Renewal*. (London: Sage, 1996), 143–144.
4. Ibid; Nicholas Garnham, "Contribution to a Political Economy of Mass Communication," in *Capitalism & Communication* (Newbury Park: Sage, 1990), 20–55.
5. Mosco, 1996, 140.
6. Eileen Meehan, "'Holy Commodity Fetish, Batman!': The Political Economy of a Commercial Intertext," in R.E. Pearson and W. Uricchio (Eds.). *The Many Lives of the Batman: Critical Approaches to a Superhero and His Media*. (New York: Routledge, 1991), 61.
7. Meehan, 1991; Janet Wasko, *Hollywood in the Information Age*. (Austin: University of Texas Press, 1995).
8. Meehan, 1991.
9. Janet Wasko, *Understanding Disney: The Manufacture of Fantasy*. (Cambridge: Polity Press, 2001); Henry Giroux, "Are Disney Movies Bad for Your Kids?" in S. Steinberg and J. Kincheloe (Eds.), *Kinderculture: The Corporate Construction of Childhood, Second Edition* (Boulder: Westview Press, 2004).
10. Naomi Klein, *No Logo: Taking Aim at the Brand Bullies*. (New York: Picador, 1999).
11. See, Wayne Munson, *All Talk: The Talkshow in Media Culture*. (Philadelphia: Temple University Press, 1993).
12. Ibid.; see also Adam Sandler, (December 23, 1996–January 5, 1997). "Warblers Warm Up at Oprah House," *Variety*, pp. 1, 58.
13. Munson, 1993. Starbucks & Oprah Fight Illiteracy (May 28, 1997). Oprah Online.

14. Steinberg and Kincheloe (1994) discuss the importance of the concept of hyper-reality to studying media culture.
15. Wasko, 1995, 214.
16. Wasko, 1995, 216.
17. Mosco, 1996.
18. Sut Jhally, (1989). "The Political Economy of Culture," In Ian Angus and Sut Jhally. (Eds.), *Cultural Politics in Contemporary America*. (New York: Routledge, 1989), 70.
19. Ibid., 77.
20. Gini Graham Scott, *Can We Talk? The Power and Influence of Talk Shows*. (New York: Insight, 1996), 252.
21. Christopher Stern, (October 30, 1995). "Backlash against TV Talk Shows," *Broadcasting & Cable*, p. 18.
22. Steve McClellan, (January 15, 1997a). "83% of GMs turned off by talk shows," *Broadcasting & Cable*. p. 3.
23. Stern, 1995, 18.
24. Joe Flint, and Gary Levin (1995). "Advertiser Won't Gamble on Trash Talk," *Variety*, November 20–26, p.17.
25. Ibid., 23.
26. Garnham, 1990, 50.
27. Ibid., 51.
28. Stern, 1995, 18.
29. McClellan, 1997a, 3.
30. Ibid.
31. Cynthia Littleton, (1996b). "The remaking of talk," *Broadcasting & Cable*. January 22, pp.46, 450.
32. Ibid.
33. Oprah's Habitat for Humanity Project. www.oprah.com, March 1997.

CHAPTER 4

1. Chip Chandler. "Self-Help Guru about to Debut on Amarillo TV," Amarillo. com. January 3, 2003. http://www.amarillonet.com/stories/010203/new_guru. shtml. Accessed on 10/3/04.
2. Taken from *Dr. Phil*'s Web site biography page: http://www.drphil.com/about/about_landing.jhtml
3. Online dictionary and references: www.xrefer.com/entry/104928.
4. Quoted in Patrick Brantlinger, *Crusoe's Footprints: Cultural Studies in Britain and America* (New York: Routledge, 1990), 96.
5. Ibid.
6. Philip Cassell, ed., *The Giddens Reader* (Stanford: Stanford University Press, 1993), 29.
7. Michel Foucault, *Language, Practice, Memory: Selected Essays and Interviews*, trans. Donald Bouchard (Ithaca: Cornell University Press, 1977), 227.
8. For a further discussion of the idea expressed here, see McLaughlin, 1993.

9. Helmut R. Wagner, ed., *Alfred Schutz: On Phenomenology and Social Relations. Selected Writings* (Chicago: The University of Chicago Press, 1970), 240.

10. For an interesting discussion on the ways in which lay knowledge and experts relate to each other, see Anthony Giddens, *Modernity and Self-Identity* (Cambridge: Polity Press, 1991).

11. Sonia Livingstone and Peter Lunt, *Talk on Television: Audience Participation and Public Debate* (New York: Routledge, 1994).

12. For detailed examples on this, see Patricia Joyner Priest, *Public Intimacies: Talk Show Participations and Tell-All TV* (Cresskill, NJ: Hampton Press, 1995).

13. From research group conducted March 11, 2001.

14. Sonia Livingstone, "Mediated Knowledge," in *Television and Common Knowledge*, ed. Jostein Gripsurd (New York: Routledge, 1999), 92–93.

15. See, for example, Jane Shattuc, *The Talking Cure: TV Talk Shows and Women* (New York: Routledge, 1997); Mimi White, *Tele-Advising: Therapeutic Discourse in American Television* (London: University of North Carolina Press, 1992); and D. Horton and R. R. Wohl "Mass Communication and Para-Social Interaction: Observations on Intimacy at a distance," in *Drama in Life: The Uses of Communication in Society*, ed., George N. Gordon (New York: Hasting House, 1976), 212–227.

16. Shattuc, *The Talking Cure: TV Talk Shows and Women*, 1997.

17. See Janice Peck, "Talk About Racism: Framing a Popular Discourse of Race on Oprah Winfrey," *Cultural Critique*, 27 (Spring 1994): 89-126.

18. C. B. Macpherson, *The Political Theory of Possessive Individualism.* (Oxford: Oxford University Press, 1962), 24–25.

19. See, Gloria Jean Masciarotte, "C'mon Girl: Oprah Winfrey and the Discourse of Feminine Talk," *Genders*, 11 (Fall 1991): 81–110; and Peck, "Talk About Racism," 1994. Lisa McLaughlin's "Chastity Criminals in the Age of Electronic Reproduction," 1993, elaborates on similar ideas. She illustrates the relation between power and expert knowledge through an interesting reading of a *Donahue* episode on "safe sex prostitution." She concludes that the discussion of the subject was framed within "restrictive binaries-male/female, madonna/whore, good/bad- and ... limited the terms of the debate to the acceptability of lifestyles and sexual practices," (52).

20. Stacy Davis and Marie-Louise Mares, "Effects of Talk Show Viewing on Adolescents," *Journal of Communication*, 48 (Summer 1998): 11.

21. Vicki Abt and Mel Seesholtz, "The Shameless World" Revisited," *Journal of Popular Film and Television*, 26. 1, (Spring 1998): 45–46.

22. "After the Show" interview, www.oprah.com.

23. Ibid. 44.

24. Shattuc, *The Talking Cure*, 10.

25. Ibid., 31.

26. Jethro K. Lieberman, *The Tyranny of Experts: How Professionals Are Closing the Open Society*, (New York: Walker and Company, 1970), 9.

27. Christopher Lasch, *The Culture of Narcissism: American Life in an Age of Diminishing Expectations* (New York: W. W. Norton, 1978).

28. The expression is from Ken Plummer, *Telling Sexual Stories: Power, Change and Social Worlds* (New York: Routledge, 1995), 19.

29. Ibid.

30. Giddens, *Modernity and Self-Identity*, 181.

31. Ibid., 34. For the deployment of therapeutic and confessional discourses on television see Mimi White. *Tele-Advising: Therapeutic discourse in American Television*, 1992.

32. Christopher Lasch, *The Culture of Narcissism*, 10.

33. Ibid., 32.

34. Ibid., 83

35. Denene Millner, "Iyanla: The New Talk of the Town," *Arts and Lifestyle: Television*. December 12, 2000.

36. V. Lyle Harris; Ibid.

37. Information on the show hosts biographies is taken from each show's home page. http://www.tvtalkshows.com.

38. From biography section on www.montelshow.com.

39. Richard Dyer, *Stars* (London: British Film Institute, 1998), 38.

40. The phrase is borrowed from Howard Spencer, quoted in Mimi White, *Tele-Advising: Therapeutic Discourse in American Television*, 29.

CHAPTER 5

1. Quoted in Wayne Munson, *All Talk: The Talkshow in Media Culture*. (Philadelphia: Temple University Press, 1993), 1.

2. For a complete review on the frequent uses of these terms in popular media, see Vicki Abt and Leonard Mustazza, *Coming After Oprah: Cultural Fallout in the Age of the TV Talk Show* (Bowling Green, OH: Bowling Green State University Popular Press, 1997), 11–12; also, Nancy Day, *Sensational TV: Trash or Journalism* (Berkeley Heights, NJ: Enslow Pubishers, 1996).

3. For an interesting study that incorporates the notion of "carnivalesque" into the study of women's reception of talk shows, see Julie Engel Manga, *Talking Trash: The Cultural Politics of Daytime TV Talk Shows* (New York: New York University Press, 2003).

4. *Jerry Springer: Too Hot for TV*, VHS, E-Realbiz.com, 1998.

5. Mikhail Bakhtin, *Rabelais and His World* (Cambridge: MIT Press, 1968).

6. Ibid., 30. Bakhtin, in fact, places his definition of the grotesque body in the much larger category of "grotesque realism" through which he reads Rabelais' work. What he means by this category is "the specific type of imagery inherent in to the culture of folk humor in all its forms and manifestations."

7. Ibid., 11–12.

8. Quoted in Peter Stallybrass and Allon White, *The Poetics and Politics of Transgression* (Ithaca, NY: Cornell University Press, 1986), 13.

9. Ibid., for an interesting and extensive review of the critical debate surrounding Bakhtin's concept of "carnival" and "transgression" see the introduction. See also, Robert Stam, "On the Carnivalesque," *Wedge*, 1 (1982): 47–55.

10. Stallybrass and White, *The Poetics and Politics of Transgression*, 26.

12. The distinction between "high" and "low" cultures and the sociopolitical implications of such an elitist perspective have been the subject of ongoing critical debates which have generated an impressive body of literature. See, for instance, Brantlinger, *Crusoe's Footprints*, 1990; Lawrence Grossberg, Cary Nelson, and Paula Treichler, eds., *Cultural Studies* (New York: Routledge, 1992); Chandra Mukerji and Michael Schudson, ed., *Rethinking Popular Culture: Contemporary Perspectives in Cultural Studies* (Berkeley, CA: University of California Press, 1991); Williams, 1958, 1961.

12. Stallybrass and Allon, *The Poetics and Politics of Transgression*, 6.

13. James Twitchell, *Carnival Culture: The Trashing of Taste in America*, (New York: Columbia University Press, 1992). In *Carnival Culture*, Twitchell makes interesting arguments about the role of the modern entertainment industries in the "trashing of taste in America." The show business, he contends, has blurred the boundaries between high art and low cultures, and reversed the "canonized" order so much so that we inhabit a world of "Lower Aesthetica" and "Upper Vulgaria." The major flaw in Twitchell's study, however, is its uncritical adoption of "upper" and "lower" categories as a naturally "given" order of things.

14. For further discussion on the cyclic nature of popular festivities, see Martin Barbero, *Communication, Culture and Hegemony: From the Media to the Mediation* (London: Sage, 1993), 88.

15. See Carpignano et al. "Chatter in the Age of Electronic Reproduction," *Social Text* 25–26, (1990): 47.

16. *Jerry Springer: Too Hot for TV*, VHS, (E-Realbiz.com, 1998). Speech reproduced from Jerry Springer's concluding notes in the *Too Hot for TV* video.

17. For an interesting discussion on the oppressive ideologies acted out in pro-wrestling, see A. D. Gresson III, "Professional Wrestling and Youth Culture: Teasing, Taunting, and the Containment of Civility," in Shirley R. Steinberg and Joe L. Kincheloe, eds., *Kinderculture: The Corporate Construction of Childhood* (Boulder, CO: Westview Press, 1997), 165–180.

18. For an interesting comparison between domestic melodrama and daytime soap operas, see Ellen Seiter, "The Promise of Melodrama" (Ph.D. diss., Northwestern University, 1981).

19. Melodrama is a theatrical production mostly popular in Europe and the Unites States in the eighteenth and nineteenth centuries. This genre has been considered, until fairly recently, a tasteless form of entertainment which catered to the "vulgar" masses of the population. The rowdiness of the audience and its complicity in the performance were demonstrated through such noisy approval and/or exuberant disapproval of the staged dramas that: "theaters had to be periodically rebuilt and redecorated as a result of the

damages caused by the public." Quoted from Martin Barbero, *Communication, Culture and Hegemony*. For an enlightened discussion on the nature and social function of melodrama, see also Peter Brooks, *The Melodramatic Imagination* (New York: Columbia University Press, 1984); and, Michael Hayes and Anastasia Nikolopoulou, *Melodrama: The Cultural Emergence of a Genre* (New York: St Martin's Press, 1996).

20. Charles Acland, *Youth, Murder, Spectacle: The Cultural Politics of "Youth in Crisis"* (San Francisco: Westview, 1995), 7.

21. See Dick Hebdige, *Hiding in the Light* (New York: Routledge, 1988), 8.

22. See Naomi Wolf, *The Cult of Beauty: How Images of Beauty Are Used Against Women* (New York: Doubleday, 1991).

23. Charles Acland, *Youth, Murder, Spectacle*, 20–25.

24. See Dick Hebdige, *Subculture and the Meaning of Style* (London: Methuen, 1979). For a discussion of youth culture and gender issues, see Angela McRobbie, *Feminism and Youth Culture* (London: Macmillan, 1991).

25. For a full discussion on the normative role of the media in solving youth crises, see Acland, *Youth, Murder, Spectacle*.

26. See, for instance, S. Bradley Greenberg, "Daytime Television Talk Shows: Guests, Content and Interactions," *Journal of Broadcasting and Electronic Media* 41. 3 (1997): 412–426.; Patricia Priest, *Public Intimacies*; and, Jane Shattuc, *The Talking Cure*.

27. For a discussion of these images, see for instance, Mary Brown, *Soap Opera and Women's Talk* (London: Sage, 1994).

28. This dichotomization between images of virgin and whore have been explored in many feminist studies. See, for instance, G. Jean Masciarotte, "C'mon Girl."

29. Mary Russo, *The Female Grotesque: Risk, Excess and Modernity* (New York: Routledge, 1994), 7.

30. Ibid., 12.

31. The tradition of "sobbing sisters" developed in the last quarter of the nineteenth century with American yellow journalism. With the rise of mass produced and mass circulated papers, the business of yellow journalism developed into a fully fledged "tabloid" industry. Female readers became among the most valued and targeted readers with emotionally charged advice columns and problem pages -hence, the sobbing sisters. For more on this, see Shattuc, *The Talking Cure*, 14–20.

CHAPTER 6

1. Michel Foucault, *The History of Sexuality*, trans. Robert Hurley (New York: Vintage Books, 1990), 40.

2. For the purposes of this chapter, the daytime television talk show is one which purports to tackle "social issues." Talk shows which predominantly feature celebrity interviews, such as *Rosie*, *Ellen*, and *Live with Regis and Kelly*, are not within the scope of inquiry. Furthermore, this essay will draw from examples

from *Maury*, *Jenny Jones* and *Ricki Lake* as these programs were easily accessible and have weekly show synopses available on the Internet. But, perhaps even more importantly, these shows combine information with spectacle in ways the *The Oprah Winfrey Show*, *Dr. Phil*, *The Ananda Lewis Show* and *Iyanla* do not. This is not to say that these programs construct female bodies and sexuality differently. We would argue that they do not; they merely contextualize them differently.

3. Here norms are understood as "expectations of how to behave shared by those in a group, culture, or society...A sense of right versus wrong applies," as quoted from Nanette J. Davis and Clarice Statz, *Social Control of Deviance: A Critical Perspective* (New York: McGraw-Hill, 1990), 4. It is also useful here to delinate between sex and gender—two terms which all too often get used for one another. Sex refers to the physical, biological designation of male or female. Gender refers to the social experience of male and female. It points to questions of identity and the ways in which people experience their sex.

4. Sandra Lee Bartky, *Feminism and Domination* (New York: Routledge, 1990), 74.

5. Michel Foucault, *The History of Sexuality*, 11.

6. While this show title does not specifically mention women, the subjects of the show are women and girls with facial deformities.

7. This episode offers updates of fidelity and paternity shows.

8. This episode tells the story of a mother who lost custody of her seven children because she parties too much.

9. This episode would also fit under appearance category since it focuses on teens that dress too provocatively.

10. This episode gives men lie detector tests to see if they are cheating on their girlfriends/wives.

11. Stacy Davis and Marie-Louise Mares, "Effects of Talk Show Viewing on Adolescents," *Journal of Communication*, 48 (Summer 1998): 84. However, with this study, one needs to consider whether the likelihood to see these issues as important is solely tied to talk show viewing or if other social, political, and economic factors may also play a part in the world outlook of these "heavy" viewers.

12. For more discussion on the construction and maintenance of stereotypes, see Homi Bhabba, "The Other Question," in *Black British Cultural Studies*, eds. Houston A. Baker, Jr. et al. (Chicago: The University of Chicago Press, 1996), 98.

13. See essays in Linda Gordon, ed., *Women, the State, and Welfare* (Madison, WI: University of Wisconsin Press, 1990) and Joan B. Landes, ed., *Feminism, the Public & the Private* (New York: Oxford University Press, 1998).

14. Stuart Hall, "Race, Articulation, and Societies Structured in Dominance," in *Black British Cultural Studies*, eds. Houston A. Baker, Jr. et al. (Chicago: The University of Chicago Press, 1996), 55.

15. bell hooks, quoted in Hazel V. Carby, "White Woman Listen," in *Black British Cultural Studies*, eds. Houston A. Baker, Jr. et al. (Chicago: The University of Chicago Press, 1996), 72.

16. Ricki Solinger, *Wake Up Little Suzie* (New York: Routledge, 2000).

17. This understanding of racism and domination is highlighted by Wahneema Lubiano when she talks about narratives of the state: "domination is so successful precisely because it sets the terrain upon which struggle occurs at the same time that it preempts opposition not only by already inhabiting the vectors where we could resist (i.e., by being powerfully in place and ready to appropriate oppositional gestures), but also by having already written the script that we have to argue within and against." Wahneema Lubiano, "Like Being Mugged by a Metaphor: Multiculturalism and State Narrative," in *Mapping Multiculturalism*, eds. Avery F. Gordon and Christopher Newfield (Minneapolis: University of Minnesota Press, 1996), 66.

18. Christopher Newfield and Avery F. Gordon, "Multiculturalism's Unfinished Business," in *Mapping Multiculturalism*, eds. Avery F. Gordon and Christopher Newfield (Minneapolis: University of Minnesota Press, 1996), 106.

19. Angela Davis, "Gender, Class and Multiculturalism: Rethinking 'Race' Politics," in *Mapping Multiculturalism*, eds. Avery F. Gordon and Christopher Newfield (Minneapolis: University of Minnesota Press, 1996), 47.

20. Wendy Brown, "Injury, Identity, Politics," in *Mapping Multiculturalism*, eds. Avery F. Gordon and Christopher Newfield (Minneapolis: University of Minnesota Press, 1996), 153.

21. And human implies a white and male body.

22. Newfield and Gordon suggest: "The culturalism of multiculturalism threatens to shift attention from racialization to culture and in so doing to treat racialized groups as one of many diverse and interesting cultures. This makes racism more difficult to acknowledge and control: phenomena that might be at least partially attributed to racism, such as crime in racially segregated, low-income neighborhoods, can be attributed in a culturalist approach to the deficient, or just different, values of the sufferers. Seeing whiteness as cultural produces a complementary blindness. While the attention to whiteness as a cultural system has been a very important move in the reconceptualization of race problems, it downplays the ongoing existence of white supremacy as a system of privilege, favoritism, discrimination, and exclusion. Indeed, given existing racial inequalities and the continuing segregation of most social institutions, the reduction of all racial groups to a nonexistent level playing field poses serious problems." Christopher Newfield and Avery F. Gordon, "Multiculturalism's Unfinished Business," 79.

23. Montel Williams, http://www.montelshow.com/montel/ashesees.htm, accessed 4–30–02.

24. Davis, "Gender, Class and Multiculturalism: Rethinking 'Race' Politics," 43.

25. Some estimates suggest 1 in 4 children are living at or near the poverty line— See Ricki Solinger, "Dependency and Choice: The Two Faces of Eve" in

Whose Welfare?, ed. Gwendolyn Mink (Ithaca, NY: Cornell University Press, 1999).

26. See Kincheloe and Steinberg, *Changing Multiculturalism*.

27. Jeanne Albronda Heaton and Nona Leigh Wilson, *Tuning in Trouble: Talk TV's Destructive Impact on Mental Health* (San Franciso: Jossy-Bass Publishers, 1995), 157.

28. Episode titled "Help! My Wild Teen Beats Me" from *Sally*, air date 8–02–00.

29. Sandra Lee Bartky, *Feminism and Domination* (New York: Routledge, 1990), 80.

30. www.spe.sony.com/tv/shows/ricki, Coming Soon, accessed 10–01–01.

31. www.spe.sony.com/tv/shows/ricki/upcoming.phtml, accessed 11–4–01.

32. Ibid.

33. www.studiousa.com/maury, Episode guide, accessed 10–08–01.

34. www.spe.sony.com/tv/shows/ricki/upcoming.phtml, accessed 10–01–01.

35. This is connected to questions of patriarchy we will discuss in the next section. However, it is interesting that on these programs, the question that is almost always asked is "Where's the father?" His presence is conspicuously absent.

36. Nanette J. Davis and Clarice Statz, *Social Control of Deviance: A Critical Perspective* (New York: McGraw-Hill, 1990), 226.

37. Ricki Solinger, *Wake Up Little Suzie*, 42.

38. Heaton and Wilson, *Tuning in Trouble: Talk TV's Destructive Impact on Mental Health*, 153–154.

39. Surprisingly, there were no "sexuality" oriented shows during the survey periods. Sexuality oriented shows feature gay and lesbian guests, cross-dressers, transvestites, and transsexuals. They are a common staple of most talk shows. For further research on gay and lesbians in talk shows see Heaton and Wilson's *Tuning in Trouble* (1995) and Joshua Gamson, *Freaks Talk Back: Tabloid Talk Shows and Sexual Nonconformity* (Chicago: University of Chicago Press, 1998).

40. Ricki Solinger, *Wake Up Little Suzie*, 9.

41. These titles come from the following programs and air dates—*Maury* 10–11–01, *Ricki Lake* 10–17–01, and *Jenny Jones* 10–15–01, respectively.

42. www.jennyjones.warnerbros.com, Weekly Show Schedule, accessed 10–14–01.

43. www.studiosusa.com/maury, Episode Guide, accessed 10–08–01.

44. www.studiosusa.com/maury, Episode Guide, accessed 12–08–01.

45. This is in contrast to the male "teen-out-of-control" who is characterized by his violence, not by his promiscuity. As such, the male "tooc" has been classified under violent masculinity.

46. From "Send My Wild Teen to Boot Camp," air date 8–03–00.

47. In a few episodes, the girls are forced to walk around a city wearing placards stating their out-of-control behavior.

48. Todd Gitlin, "Prime Time Ideology: The Hegemonic Process in Television Entertainment," in *Television: A Critical View*, 3rd edition, ed. Horace Newcomb (New York: Oxford University Press, 1987), 510. emphasis Gitlin.

49. www.sallyjr.com/sally4/frm_bootcamps.html, accessed 10–01–01.

50. Meda Chesney-Lind, "Women and Crime: The Female Offender," in *Gender, Crime and Feminism*, ed. Ngaire Naffine (Brookfield, VT: Dartmouth, 1995), 8.

51. Ibid., 3–22.

52. Ibid., 21. Further support and evidence indicating women's punishment for moral offenses is given in David Downes and Paul Rock, *Understanding Deviance*, 3rd edition (New York: Oxford, 1998). These authors write, "Adolescent girls face much higher risks of institutionalization than boys for non-criminal forms of sexual 'deviance.'" They suggest the cause of this is that "unreconstructed notions about women's 'nature' have lent undue prominence to 'sexual deviance' as the focus of inquiry appropriate to the study of female criminality." However, the question remains as to where these 'unreconstructed notions' come from. We can say that women are punished for moral offenses because of notions about women's nature, but that doesn't seem to really get at the heart of it. Simply put, why isn't it ok for women to be publicly sexual? We must admit we are leery of this question because initially it points to biological answers, seemingly a deadend. Beyond the ascribed nurturer role, there is the notion of paternity—a man's claim to a child. No one has come out and said this though, just 'notions about women's nature.' This notion of women's nature is also reinforced in Clyde Vedder and Dora Somerville, *The Delinquent Girl* (Springfield, IL: Charles C. Thomas Publishing, 1975). This book was written by a criminologist and a social worker as a guide for social workers, police officers, teachers, etc. in handling and defining the delinquent girl. The book, written in 1975, is mostly outdated; however, it is relevant because it frames the context of the construction and development of institutional knowledges about the social roles of women. The authors insist that the natural inclination of females is to be a nurturer and caregiver. They argued that it was the job of social workers to resocialize delinquent girls to take her role in the traditional family. This was suggested by giving the girls access to classes in home economics—sewing, cooking, etc. Although social roles for women have changed since the days of Vedder and Sommerville, they are still articulated through the development of professional knowledges, which focus on the individual woman's body.

53. Kenneth Thompson, *Moral Panics* (New York: Routledge, 1998), 85.

54. Thompson illustrates this idea through the notion of illegitimacy and family in Britain where conservative politicians used the notion of illegitimate children as the cause of the moral and social decline of the traditional community and increase in crime. Ricki Solinger makes a similar case in the U.S. She argues that black unwed mothers are portrayed as highly sexual, promiscuous breeders, and women who have children to get more money from the welfare system. On the other hand, white unwed mothers are seen as mentally ill, but able to be rehabilitated. In this context, black illegitimate children are seen as drains on society. White illegitimate children are seen as valuable

commodities on the adoption market. As such, blackness gets positioned with immorality. Thompson and others suggest instead that we look at illegitimacy as a symptom rather than the locus of a problem.

55. Kerry Carrington, "Feminist Readings of Female Delinquency," in *Gender, Crime and Feminism*, ed. Ngaire Naffine (Brookfield, VT: Dartmouth, 1995), 140.

56. Nancy Fraser, "Struggle Over Needs: Outline of a Socialist-Feminist Critical Theory of Late-Capitalist Political Culture," *Women, the State and Welfare*, ed. Linda Gordon (Madison, WI: The University of Wisconsin Press, 1990), 212.

57. Deborah Tolman, "Adolescent Girls' Sexuality: Debunking the Myth of the Urban Girl," in *Urban Girls: Resisting Stereotypes, Creating Identities*, ed. Bonnie J. Ross Leadbetter and Niobe Way (New York: New York University Press, 1996), 259.

58. Davis and Statz, *Social Control of Deviance: A Critical Perspective*, 224. Emphasis mine.

59. Sandra Lee Bartky, *Feminism and Domination*, 80. Emphasis Bartky.

60. Michel Foucault, *Discipline and Punish*, trans. Alan Sheridan (New York: Vintage, 1995), 25.

61. Judith Allen, "Men, Crime and Criminology: Recasting the Questions," in *Gender, Crime and Feminism*, ed. Ngaire Naffine (Brookfield, VT: Dartmouth, 1995), 99–120.

62. Davis and Statz, *Social Control of Deviance: A Critical Perspective*, 232. In more recent times, the recent reinstitution of virginity testing in South Africa as a means of deterring the AIDS epidemic is alarming. The logic is that by monitoring girls' virginity, they will be less likely to have sex and bring shame on themselves and their families. While this testing is not mandated by the state, girls are often pressured or forced into the testing by their families. The virginity testing requires a pelvic examination, and if a girl passes, she gets a stamp of approval. There is no equivalent test for boys. What is perhaps even scarier (for lack of a better word) is that virginity testing is being done in places such as schools and community centers where the girls receive their badge and then go home. In a country where there is a widely held superstition that sleeping with a virgin will cure AIDS, virginity testing puts these girls at grave risk of rape. Turkey is also considering instating virginity testing.

63. James Messerschmidt, "Feminism, criminology and the rise of the female sex 'delinquent', 1880–1930," in *Gender, Crime and Feminism*, ed. Ngaire Naffine (Brookfield, VT: Dartmouth, 1995), 51–72.

64. Ibid., 51.

65. Ibid., 53.

66. Lorraine Gelsthorpe, "Towards a Sceptical Look at Sexism," in *Gender, Crime and Feminism*, ed. Ngaire Naffine (Brookfield, VT: Dartmouth, 1995), 23–50.

67. Tolman, "Adolescent Girls' Sexuality: Debunking the Myth of the Urban Girl," 255.

68. Michel Foucault, *The History of Sexuality*, 33.

CHAPTER 7

1. From "Paternity Questions. . .Are the Rumors True?" *Montel*. CBS. WUSA, Washington, DC May 6, 2002.

2. Andrea Sachs, "Can They De-Springerize Talk? These hosts say viewers want a new alternative to 'screamer' shows. Oprah, meet Iyanla and Ananda," *Time*, August 6, 2001, 56.

3. From "Paternity Questions. . .Are the Rumors True?" *Montel*. CBS. WUSA, Washington, DC May 6, 2002.

4. From "The Wildest, Most Memorable *Maury* Guests" *Maury*. Fox. WTTG, Washington, DC May 6, 2002.

5. From "Who Is the Father of My Baby" *Jenny Jones*. Fox. WTTG, Washington, D.C. 30 July 2002.

6. It is worth noting that guests like Tiffany provide talk shows with content for future shows, both through retesting and update shows. In a few instances, some guests appear three or more times.

7. Rosemary Hennessy, *Profit and Pleasure: Sexual Identities in Late Capitalism* (New York: Routledge, 2000), 23.

8. We understand that these are rather substantial generalizations. We by no means want to diminish fathers and recognize that there may be individual cases that do not gel with what we have presented. However, we are speaking of paternity cases, where a parent, either mother or father, is petitioning the court. When the mother is petitioning the court, she is more than likely seeking, at the very least, child support payments. When the father is petitioning the court, he may be looking for custody, visitation, reduced child support payments, or to disprove paternity, thereby eliminating child support payments.

9. Barbara Katz Rothman, "Beyond Mothers and Fathers: Ideology in a Patriarchal Society," in *Mothering: Ideology, Experience, and Agency*, eds. Evelyn Nakano Glenn, et al. (New York: Routledge, 1994), 139.

10. Ruth Hubbard, *Profitable Promises* (Monroe, ME: Common Courage Press, 1995), 179.

11. See Jose Van Dijck, *Imagenation: Popular Images of Genetics* (New York: New York University Press, 1998); Ruth Hubbard, *Profitable Promises*; Dorothy Nelkin and M. Susan Lindee, *The DNA Mystique: The Gene as a Cultural Icon* (New York: Freeman, 1995); and Roger Lancaster, *The Trouble with Nature: Sex and Science in Popular Culture* (Berkeley: University of California Press, 2003)—to name a few.

12. Patricia Hill Collins, "Shifting the Center: Race, Class, and Feminist Theorizing About Motherhood," in *Mothering: Ideology, Experience, and Agency*, eds. Evelyn Nakano Glenn, et al. (New York: Routledge, 1994), 63.

13. Ibid., 63.
14. Ruth Hubbard, *Profitable Promises*, 189. It is worth noting that the term "nature" is also contentious.
15. Ruth Hubbard and Elijah Wald, *Exploding the Gene Myth: How Genetic Information Is Produced and Manipulated by Scientists, Physicians, Employers, Insurance Companies, Educators, and Law Enforcers* (Boston: Beacon Press, 1993), 7.
16. Ruth Hubbard, *Profitable Promises*, 169–176. In each of these instances the "female" component of the pair is seen as stabile, passive, fixed. It is acted upon by the male component in the pair. The resulting child is a product of the male's action.
17. Marita Sturken and Lisa Cartwright, *Practices of Looking: An Introduction to Visual Culture* (New York: Oxford University Press, 2001), 299.
18. From "The Wildest, Most Memorable *Maury* Guests" *Maury*. Fox. WTTG, Washington, DC May 6, 2002.
19. From "Angry Teen Moms Demand Paternity Tests" *Maury*, Fox WTTG, Washington, DC February 1, 2001
20. From "Who Is the Father of My Baby" *Jenny Jones*. Fox. WTTG, Washington, DC July 30, 2002.
21. E. Ann Kaplan, "Look Whose Talking, Indeed: Fetal Images in Recent North American Culture," in *Mothering: Ideology, Experience, and Agency*, eds. Evelyn Nakano Glenn, et al. (New York: Routledge, 1994), 123.
22. Ibid., 132.
23. For a phenomenological understanding of the connection between mothers and children see Susan Bordo, "Are Mothers Persons? Reproductive Rights and the Politics of Subject-ivity," in *Unbearable Weight* (Berkeley: University of California Press, 1993).
24. Sturken and Cartwright, *Practices of Looking: An Introduction to Visual Culture*, 299.
25. It is important to note that the moment at which a woman becomes a mother is hotly contested. In this instance, we are speaking biologically. We are aware of the debates around reproductive rights and abortion issues. However, these debates are not the focus here. Regardless of belief in that arena, once the decision has been made to label pregnant woman as mother, the mother is then expected to fulfill certain roles, duties, and responsibilities.
26. Sturken and Cartwright, *Practices of Looking: An Introduction to Visual Culture*, 286.
27. See Ruth Hubbard and Elijah Wald's *Exploding the Gene Myth* and Ruth Hubbard's *Profitable Promises* for a debunking of some of the prophetic promises of genetic testing. See Jose Van Dijck's *Imagenation* for an account of the gene debate as portrayed in the popular media.
28. For instance, on April 16, 2003, the National Institutes of Health held a symposium titled "Genes, Brain, Behavior: Before and Beyond Genomics." Featured panels included one on Learning and Memory and another on

alcoholism. This symposium was funded in part by the Department of Energy, the Department of Health and Human Services, Merck & Co., Inc, GlaxoSmithKlein and IBM Life Sciences.

29. A search of the Computer Retreval of Information and Scientific Projects (crisp.cit.nih.gov, 4–11–03) reveals only five studies on paternity—three of which were seeking to make paternity testing cheaper.

30. According to the 2000 U.S. Census, "One parent familes maintained by women were also more likely than those maintained by men to have family incomes below the poverty level (34 percent compared with 16 percent)." From Jason Fields and Lynne M. Casper, *America's Families and Living Arrangements: March 2000*, Current Population Reports: P20–537 (U.S. Census Bureau, Washington, DC, 2001), 8.

31. Alternatively, in some cases these rights and responsibilities are not terminated until he is proven not to be the father.

32. David Popenoe, "Life Without Father," in *Lost Fathers: The Politics of Fatherlessness in America*, ed. Cynthia R. Daniels (New York: St. Martin's Press, 1998), 36.

33. Recent developments enable testing facilities to test without the mother. However, it is preferred to have the mother present.

34. Eric Dexheimer, "Pop Quiz: As Genetic Testing Pins Down Paternity Issues, the Definition of Fatherhood gets Murkier," *Westwood*, June 7, 2001, 16.

35. Ibid.

36. Fairfax Identity Laboratories. http://www.fairfaxidlab.com. Date accessed 28 August 2004.

37. DNA Diagnostic Center. http://www.dnacenter.com/media.html. Date accessed August 28, 2004.

38. Irwin Garfinkel, *Assuring Child Support: An Extension of Social Security* (New York: Russell Sage Foundation, 1992), 70.

39. According to the 2000 U.S. Census, there are four to five times as many female single-parent households than male, depending on the age of the children. Furthermore, there are over 10 times as many female single-parent households living below poverty than there are male households. And, a disproportionate number of those families living below poverty are black families. From Jason Fields and Lynne M. Casper. 2001. *America's Families and Living Arrangements: March 2000*. Current Population Reports: P20–537. (U.S. Census Bureau, Washington, DC), 8.

40. Jakie Boggis, quoted in "Critics Blame Child-Support System for Encouraging Fatherlessness." The CQ Researcher. 2 June, 2000, 34–36, Online. Available at http://library.cqpress.com, accessed 4–14–03.

41. David Popenoe, "Life Without Father," 47.

42. For a considered and thoughtful, if mainstream, examination of the father's rights movement, see *The CQ Researcher's* June 2, 2000, issue devoted to the Fatherhood movement.

43. Drucilla Cornell, "Fatherhood and Its Discontents: Men, Patriarchy, and Freedom," in *Lost Fathers: The Politics of Fatherlessness in America*, ed. Cynthia R. Daniels (New York: St. Martin's Press, 1998), 187. Emphasis Cornell.

44. Ibid., 198. This idea is further supported in the work of Susan Bordo and Simone de Beauvoir.

45. Historian Robert Griswold explores the history and politics of fatherlessness. In his study, he also looks at the father's movement and discusses how citizenship is conceived of as economic independence from the state. Nuclear families produce good citizens who only take from the state what they put into the state in the form of taxes. Following this logic, Griswold finds that "[i]f women do not have husbands and/or well-paid jobs, they cannot be truly independent of the state, and if they cannot be independent of the state, they are parasites—the not 'truly needy'—who too often rear disruptive, even criminally inclined children who threaten civil order." Robert L. Griswold, "The History and Politics of Fatherlessness," in *Lost Fathers: The Politics of Fatherlessness in America*, ed. Cynthia R. Daniels (New York: St. Martin's Press, 1998), 22.

46. Rosemary Hennessy, *Profit and Pleasure: Sexual Identities in Late Capitalism*, 25.

47. Title refers to an episode title of *Montel*, May 10, 2002.

48. From "I'm only in the 8th grade. . .are you my baby's daddy?" *Maury*. Fox. WTTG, Washington, DC

49. From "Angry Teen Moms Demand Paternity Tests" *Maury*. Fox. WTTG, Washington, DC February 1, 2001.

50. From "I'm only in the 8th grade. . .are you my baby's daddy?" *Maury*. Fox. WTTG, Washington, D.C.

51. Ibid.

52. From "The Wildest, Most Memorable *Maury* Guests" *Maury*. Fox. WTTG, Washington, D.C. May 6, 2002.

53. These reactions are often a response to the verbal assaults issued earlier in the program. There are few experiences as demeaning as being called a "bitch," "whore" or "scandalous troll" on national television (and in some instances, even after paternity test results in their favor, the women are still debased and called names, e.g., "you're still a whore.")

54. From "Paternity Questions. . .Are the Rumors True?" *Montel*. CBS. WUSA, Washington, DC May 6, 2002.

55. From "I'm only in the 8th grade. . .are you my baby's daddy?" *Maury*. Fox. WTTG, Washington, DC

56. At these moments, one cannot help but draw connections to the welfare system and the George W. Bush administration's support and push for the Temporary Assistance for Needy Families marriage support clause. In fact, the Bush administration, taking up fathers' rights rhetoric, proposes to spend hundreds of millions of dollars "on developing innovative approaches to supporting family formation and healthy marriages" ("Working for

Independence: Promoting Child Well-Being and Healthy Marriages." http://www.whitehouse.gov/news/releases/2002/02/welfare-book-05.html, accessed 4–15–03.) The aim is to reduce the welfare rolls by encouraging marriage, but this "encouragement" really means reducing benefits to unmarried mothers and reaffirming the traditional marriage form. For instance, according to the NOW Legal Defense and Education Fund, monies are being diverted from child support enforcement programs to marriage promotion programs, the logic being that if the couple gets married, child support won't be necessary. ("Marriage Promotion: What the Administration is Already Doing." http://www.nowldef.org/html/issues/wel/WhatAlready.pdf, accessed 4–15–03.)

57. Ricki Solinger, *Wake Up Little Suzie*, 9.
58. See also Ricki Solinger, *Wake Up Little Suzie*.
59. Ruth Hubbard, *Profitable Promises*, 64.
60. KathyKoch, "Fatherhood Movement." The CQ Researcher Online 10. 21 (June 2, 2000): 473–496. http://library.cqpress.com/cqresearcher/cqresrre2000060200 (accessed October 4, 2004).
61. "The Negro Family: The Case for National Action." Office of Policy Planning and Research, U.S. Department of Labor. Online. Accessed 4–15–03.
62. Koch, "Fatherhood Movement," 473–496.

CHAPTER 8

1. Steinberg and Kincheloe, *Kinderculture: The Corporate Construction of Childhood*.

REFERENCES

ABCNEWS.com (http:///204.202.137.115/sections/business/ DailyNews/). CBS Buying King World: $2.5 Billion Stock Deal Marries Network, Distributor. The Associated Press.

Abt, Vicki and Leonard Mustazza. (1997). *Coming after Oprah: cultural fallout in the age of the TV talk show*. Bowling Green, OH: Bowling Green State University Popular Press.

Abt, Vicki and Mel Seesholtz. (1994). "The Shameless World of Phil, Sally and Oprah: Television Talk Shows and the Deconstructing of Society." *Journal of Popular Culture*, 28 (Summer): 195–216.

Abt Vicki and Mel Seesholtz. (1998). "The Shameless World" Revisited." *Journal of Popular Film and Television*, 26, 1 (Spring): 45–46.

Acland, Charles. (1995). *Youth, Murder, Spectacle: The Cultural Politics of "Youth in Crisis*. San Francisco: Westview.

Albarran, Alan. (2002). *Media Economics.* Second edition. Iowa City: Iowa State University Press.

Allen, Judith. (1995). "Men, Crime and Criminology: Recasting the Questions." In *Gender, Crime and Feminism*. Ed. Ngaire Naffine. Brookfield, VT: Dartmouth, pp. 99–120.

"Angry Teen Moms Demand Paternity Tests" Maury. Fox. WTTG, Washington, DC February 1, 2001.

Bagdikian, Ben. (2000). *The Media Monopoly*. 6th edition. Boston: Beacon Press.

Bakhtin, M. (1968). *Rabelais and His World*. Cambridge: MIT Press.

Barbero, Martin. (1993). *Communication, Culture and Hegemony: From the Media to the Mediation*. London: Sage.

Barnouw, Erik. (1990). *Tube of Plenty: The Evolution of American Television*. 2nd rev. ed. Oxford: Oxford University Press.

Bartky, Sandra Lee. (1990). *Feminism and Domination*. New York: Routledge.

Baudrillard, J. (1983). "The Ecstasy of Communication." In *The Anti-Aesthetic*. Ed. Hal Foster. Washington: Bay Press.

Baudrillard, J. (1984). "The Procession of Simulacrum," in *Art After Modernism*. Wallis B. Ed. New York: The New Museum of Contemporary Art Publication, pp. 253–283.

Beauvoir, Simone de. (1989). *The Second Sex*. New York. Vintage Books.

Bettig, Ronald. (1996). *Copyrighting Culture: The Political Economy of Intellectual Property*. Boulder: Westview.

Bettig, Ronald and Jeanne Hall. (2003). *Big Media, Big Money: Cultural Texts and Political Economics*. Lanham, MD: Rowman & Littlefield.

Bhabba, Homi. (1996). "The Other Question." *Black British Cultural Studies*. Eds. Houston A. Baker, Jr., et al. Chicago: The University of Chicago Press, pp. 87–106.

Bogart, Leo. (1998). "What Does It All Mean?" In *Media Mergers*. Eds. Nancy J. Woodhull and Robert W. Snyder. New Brunswick/London: Transaction Publishers, pp. 17–27.

Boggis, Jackie. (2000, June 2). In "Critics Blame Child-Support System for Encouraging Fatherlessness." *The CQ Researcher*, 34–36.

Bordo, Susan. (1993). *Unbearable Weight*. Berkeley: University of California Press.

Brantlinger, P. (1990). *Crusoe's Footprints: Cultural Studies in Britain and America*. New York: Routledge.

Brooks, Peter. (1984). *The Melodramatic Imagination*. New York: Columbia University Press.

Brown, Mary. (1994). *Soap Opera and Women's Talk*. London: Sage.

Brown, Wendy. (1996). "Injury, Identity, Politics." *Mapping Multiculturalism*. Eds. Avery F. Gordon and Christopher Newfield. Minneapolis: University of Minnesota Press, pp. 166.

Burghart, Tara. (2004). *Marketing Execs Eye "Oprah" Giveaway*. Netscape Network 2004 [cited September 23, 2004]. Available from http://cnn.netscape.cnn.com/ns/news/story.jsp?flok=FF-APO-1333&idq=/ff/story/0001%2F20040914%2F2100437567.htm&sc=1333&photoid=20040902WXS104.

Carbaugh, D. (1988). *Talking American: Cultural Discourse on Donahue*. New Jersey: Alex Publishing Corporation.

Carby, Hazel V. (1996). "White Woman Listen! Black Feminism and the Boundaries of Sisterhood." In *Black British Cultural Studies*. Eds. Houston A. Baker, Jr., et al. Chicago: The University of Chicago Press, pp. 61–86.

Carpignano, P., R. Andersen, S. Aronowitz, and W. Difazio. (1990). "Chatter in the Age of Electronic Reproduction: Talk Television and the "Public Mind." *Social Text*, (25–26): 33–55.

Carrington, Kerry. (1995). "Feminist Readings of Female Delinquency." In *Gender, Crime and Feminism*. Ed. Ngaire Naffine. Brookfield, VT: Dartmouth, pp. 121–148.

Cassell, Philip. (Ed.). (1993). *The Giddens Reader*. Stanford University Press.

"CBS Buys King World." E! Online News Staff. April 1, 1999 http://www.eonline.com/News?Items?0%2C1%2C4553%2C00%2Ehtml

Chandler, Chip. "Self-Help Guru about to Debut on Amarillo TV." Amarillo.com, January 3, 2003. http://www.amarillonet.com/stories/010203/new_guru.shtml. Accessed on 10/3/04.

Chesney-Lind, Meda. (1995). "Women and Crime: The Female Offender." In *Gender, Crime and Feminism*. Ed. Ngaire Naffine. Brookfield, VT: Dartmouth, 3–22.

Cohen, Jeff (May 4, 1997). "TV Industry Wields Power in D.C." *The Baltimore Sun*.

Collins, Patricia Hill. (1994). "Shifting the Center: Race, Class, and Feminist Theorizing about Motherhood." In *Mothering: Ideology, Experience, and Agency*. Eds. Evelyn Nakano Glenn, et al., New York: Routledge, pp. 45–65.

Cornell, Drucilla. (1998). "Fatherhood and Its Discontents: Men, Patriarchy, and Freedom." In *Lost Fathers: The Politics of Fatherlessness in America*. Ed. Cynthia R. Daniels. New York: St. Martin's Press, pp. 183–202.

The CQ Researcher Online http://library.cqpress.com/cqresearcher/cqresrre2000060200.

Davis, Angela. (1996). "Gender, Class and Multiculturalism: Rethinking "Race" Politics." In *Mapping Multiculturalism*. Eds. Avery F. Gordon and Christopher Newfield. Minneapolis: University of Minnesota Press, pp. 40–48.

Davis, Nanette J. and Clarice Statz. (1990). *Social Control of Deviance: A Critical Perspective*. New York: McGraw-Hill.

Davis, Stacy and Marie-Louise Mares. (1998). "Effects of Talk Show Viewing on Adolescents." *Journal of Communication*, 48 (Summer): 84.

Day, Nancy. (1996). *Sensational TV: Trash or Journalism*. Berkeley Heights, NJ: Enslow Publishers.

Debord, G. (1990). *Comments on the Society of the Spectacle*. Trans. Imrie M. New York: Verso.

Debord, G. (1967). *La Societe du Spectacle*. Paris: Gallimard.

Dexheimer, Eric. "Pop Quiz: As Genetic Testing Pins down Paternity Issues, the Definition of Fatherhood Gets Murkier." *Westwood*, June 7, 2001, 16.

DNA Diagnostic Center. http://www.dnacenter.com/media.html.

Domhoff, William. (2005). *Who Rules America? Power, Politics, and Social Change*. New York: McGraw Hill

Downes, David and Paul Rock. (1998). *Understanding Deviance*. 3rd edition. New York: Oxford University Press.

Dyer, Richard. (1998). *Stars*. London: British Film Institute.

Eco, U. (1985). "The Frame of the Comic 'Freedom'." In *Carnival*. Ed. Thomas A. Seboek. Berlin: Mouton.

Ehrenreich, Barbara (1995 December 4). "In Defense of Talk Shows." *Time*. p. 92.

Ewen, Stuart. (1976). *Captains of Consciousness: Advertising and the Social Roots of the Consumer Culture*. New York: McGraw-Hill.

FAIR. "The Government just Gave the Commercial Broadcasters More Than $50 Billion of Your Property. And All You Got Was a Bigger Wasteland," *FAIR*. New York: FAIR.

Fairfax Identity Laboratories. http://www.fairfaxidlab.com. Date accessed August 28, 2004.

Fiske, John. (1987). *Television Culture*. London: Routledge.

Fields, Jason and Lynne M. Casper. (2001). America's Families and Living Arrangements: March 2000. *Current Population Reports*: P20–537. U.S. Census Bureau, Washington, DC.

Flint, Joe. (1995a September 4–10). "NBC links with Par, P&G team," *Variety*, p. 23.

Flint, Joe (1995b September 4–10). "Talk is cheap, but lucrative," *Variety*, p. 38.

Flint, Joe and Gary Levin. (1995 November 20–26). "Advertiser Won't Gamble on Trash Talk," *Variety*, pp. 17, 23.

Foucault, Michel. (1995). *Discipline and Punish*. Trans. Alan Sheridan. New York: Vintage.

Foucault, Michel. (1990). *The History of Sexuality*. Trans. Robert Hurley. New York: Vintage Books.

Foucault, Michel. (1977). *Language, Practice, Memory: Selected Essays and Interviews*. Trans. Donald Bouchard. Ithaca: Cornell University Press.

Fraser, Nancy. (1990). "Struggle over Needs: Outline of a Socialist-Feminist Critical Theory of Late-Capitalist Political Culture." In *Women, the State and Welfare*. Ed. Linda Gordon. Madison, WI: The University of Wisconsin Press.

Gamson, Joshua. (1998). *Freaks Talk Back: Tabloid Talk Shows and Sexual Nonconformity*. Chicago: University of Chicago Press.

Garfinkel, Irwin. (1992). *Assuring Child Support: An Extension of Social Security*. New York: Russell Sage Foundation.

Garnham, Nicholas. (1990). "Contribution to a Political Economy of Mass Communication." In *Capitalism & Communication*. Newbury Park: Sage, pp.20–55.

Gelsthorpe, Lorraine. (1995). "Towards a Skeptical Look at Sexism." In *Gender, Crime and Feminism*. Ed. Ngaire Naffine. Brookfield, VT: Dartmouth University Press, pp. 23–50.

Giddens, Anthony. (1979). "Central Problems in Social Theory: Action, Structure and Contradiction." In *Social Analysis*. Berkeley: University of California Press.

Giddens, Anthony. (1984). *The Constitution of Society: Outline of the Theory of Structuration*. Berkeley: University of California Press.

Giddens, Anthony. (1990). *The Consequences of Modernity*. Stanford: Stanford University Press.

Giddens, Anthony. (1991). *Modernity and Self-Identity: Self and Society in the Late Modern Age*. Polity Press.

Giroux, Henry. (1994). *Disturbing pleasures: Learning Popular Culture*. New York: Routledge, 1994.

Giroux, Henry. (2004). "Are Disney Movies Bad for Your Kids?" In *Kinderculture: The Corporate Construction of Childhood*, Eds. S. Steinberg and J. Kincheloe. 2nd ed. Boulder: Westview Press.

Gitlin, Todd. (1987). "Prime Time Ideology: The Hegemonic Process in Television Entertainment." In *Television: A Critical View*. 3rd Edition. Ed. Horace Newcomb. New York: Oxford University Press, pp. 507–532.

Gordon, Linda (Ed). (1990). *Women, the State, and Welfare*. Madison, WI: University of Wisconsin Press.

Greenberg, S. Bradley. (1997). "Daytime Television Talk Shows: Guests, Content and Interactions." *Journal of Broadcasting and Electronic Media*, 41 (3): 412–426.

Grindstaff, Laura. (2002). *The Money Shot: Trash, Class, and the Making of TV Talk Shows*. Chicago: University of Chicago Press.

Griswold, Robert L. (1998). "The History and Politics of Fatherlessness." *Lost Fathers: The Politics of Fatherlessness in America*. Ed. Cynthia R. Daniels. New York: St. Martin's Press, pp. 11–32.

Grossberg, L., C. Nelson, and P. Treichler, (1992). *Cultural Studies*. New York: Routledge.

Haag, L. L. (1993). "Oprah Winfrey: The Construction of Intimacy in the Talk Show Setting." *Journal of Popular Culture*, 26(4): 115–121.

Hall, Stuart. (1996). "Race, Articulation, and Societies Structured in Dominance." In *Black British Cultural Studies*. Eds. Houston A. Baker, Jr., et al. Chicago: The University of Chicago Press, pp. 16–60.

Hall, Stuart (1987). "Encoding/decoding." In *Culture, Media, Language*. Eds., Stuart Hall, D. Hebson et al. London: Hutchinson.

Harris, V. Lyle. (2001). "Watch Out, Oprah: TV Audiences Are Getting Another Candidate for Talk-Show Superstardom." *Cox News Service.*

Hayes, Michael and Anastasia Nikolopoulou. (1996). *Melodrama: The Cultural Emergence of a Genre.* New York: St Martin's Press.

Heaton, Jeanne Albronda and Nona Leigh Wilson. (1995). *Tuning in Trouble: Talk TV's Destructive Impact on Mental Health.* San Francisco: Jossy-Bass Publishers.

Hebdige, D. (1988). *Hiding in the Light.* New York and London: Routledge.

Hebdige, D. (1979). *Subculture and the Meaning of Style.* London: Methuen.

"Help! My Wild Teen Beats Me" from *Sally*, air date August 2, 2000.

Hennessy, Rosemary. (2000). *Profit and Pleasure: Sexual Identities in Late Capitalism.* New York: Routledge.

Horton D. and R. R. Wohl. (1976), "Mass Communication and Para-Social Interaction: Observations on Intimacy at a Distance." In *Drama in Life: The Uses of Communication in Society.* Ed. George N. Gordon. New York: Hasting House, pp. 212–227.

Hubbard, Ruth. (1995). *Profitable Promises.* Monroe, ME: Common Courage Press,

Hubbard, Ruth and Elijah Wald. (1993). *Exploding the Gene Myth: How Genetic Information Is Produced and Manipulated by Scientists, Physicians, Employers, Insurance Companies, Educators, and Law Enforcers.* Boston: Beacon Press.

"I'm only in the 8th Grade. Are you my baby's daddy?" *Maury.* Fox. WTTG, Washington, D.C.

Iyengar, Shanto, Mark D. Peters, and Donald R. Kinder (1982). "Experimental Demonstrations of the 'not so' minimal' Consequences of Television News Programs." *American Political Science Review*, Vol. 76, December. No. 4: 848–858.

Jameson, F. (1984). "Postmodernism, or the Cultural Logic of Late Capitalism." *New Left Review*, (146): 53–92.

Jameson, F. (1983) "Postmodernism and Consumer Society." In *The Anti-Aesthetic.* Ed. Foster Hal. Washington: Bay Press, pp. 115–125.

Jerry Springer: Too Hot for TV, VHS, E-Realbiz.com, 1998.

Jhally, Sut. (1989). "The Political Economy of Culture," In *Cultural Politics in Contemporary America*. Eds., Ian Angus & Sut Jhally. New York: Routledge.

Jhally, Sut and Justin Lewis. (1992). *Enlightened Racism: The Cosby Show, Audiences, and the Myth of the American Dream*. Boulder: Westview.

Kaplan, E. Ann. (1994). "Look Who's Talking, Indeed: Fetal Images in Recent North American Visual Culture." In *Mothering: Ideology, Experience, and Agency*. Eds. Evelyn Nakano Glenn, et al. New York: Routledge, pp. 121–137.

Kaplan, J. (1995). "Are Talk Shows out of Control?" *TV Guide*, 1 (April): 10–15.

Kellner, Douglas. (1989). "Techno-Capitalism," In *Critical Theory, Marxism, & Modernity*. Baltimore: Johns Hopkins University Press.

Kellner, Douglas. (1990). *Television and the Crisis of Democracy*. Boulder: Westview.

Kellner, Douglas. (1995). *Media Culture: Cultural Studies, Identity, and Politics Between the Modern and the Postmodern*. New York: Routledge, 1995.

Kennickell. Arthur B. (2003). "A Rolling Tide: Changes in the Distribution of Wealth in the U.S., 1989–2001." Table 10. Levy Economics Institute.

Kincheloe, Joe. (2004). *Critical Pedagogy: A Primer*. New York: Peter Lang.

Kincheloe, Joe and Shirley Steinberg. (1997). *Changing Multiculturalism*. London: Open University Press.

Klein, Naomi. (1999). *No Logo: Taking Aim at the Brand Bullies*. New York: Picador.

Koch, Kathy. (2000 June 2). "Fatherhood Movement." *The CQ Researcher Online* 10. 21, 473–496.

Lancaster, Roger. (2003). *The Trouble with Nature: Sex in Science and Popular Culture*. Berkeley: University of California Press.

Landes, Joan. (1998). *Feminism, the Public and Private*. New York: Oxford University Press.

Lasch, Christopher. (1978). *The Culture of Narcissism: American Life in an Age of Diminishing Expectations*. New York: W. W. Norton.

Lieberman, Jethro K. (1970). *The Tyranny of Experts: How Professionals are Closing the Open Society.* New York: Walker and Company.

Littleton, Cynthia. (1996a January 22). "Phil Departing; Maury Returns," *Broadcasting & Cable*, p. 119.

Littleton, Cynthia. (1996b January 22). "The remaking of talk," *Broadcasting & Cable*, pp. 46, 450.

Livingstone, Sonia. (1999). "Mediated Knowledge." In *Television and Common Knowledge.* Ed. Jostein Gripsurd. London and New York: Routledge, pp. 91–107.

Livingstone, Sonia and Peter Lunt. (1994). *Talk on Television: Audience Participation and Public Debate.* New York: Routledge.

Lubiano, Wahneema. (1996). "Like Being Mugged by a Metaphor: Multiculturalism and State Narrative." In *Mapping Multiculturalism.* Eds. Avery F. Gordon and Christopher Newfield Minneapolis: University of Minnesota Press, pp. 64–75.

MacEwan, Arthur. (1998 May/June). "Dr. Dollar." *Dollars & Sense.*

Macpherson, C.B. (1962). *The Political Theory of Possessive Individualism.* Oxford: Oxford University Press.

Manga, Julie Engel. (2003). *Talking Trash: The Cultural Politics of Daytime TV Talk Shows.* New York: New York University Press.

Marriage Promotion: What the Administration Is Already Doing. NOW Legal Defense and Education Fund. http://www.nowldef. org/html/issues/wel/WhatAlready.pdf, accessed 4–15–03.

Martin-Barbero, J. (1993). *Communication, Culture and Hegemony: From the Media to the Mediation.* London: Sage.

Marx, Karl. (1978). "Capital: Volume One," In *The Marx-Engels Reader*, 2nd ed. Ed., Robert C. Tucker. New York: Norton. p. 321.

Masciarotte, Gloria-Jean. (1991). "C'mon Girl: Oprah Winfrey and the Discourse of Feminine Talk." *Genders*, 11 (Fall): 81–110.

McChesney, Robert W. (1996 July). "The Global Struggle for Democratic Communication." *Monthly Review.* 48.3, 1–20.

McChesney, Robert W. (1997a). *Corporate Media and the Threat to Democracy.* New York: Seven Stories Press.

McChesney, Robert W. (1997b April 21). "Digital Highway Robbery." *Nation.* 264.15, 22–24.

McClellan, Steve. (1997a January 15). "83% of GMs Turned off by Talk Shows," *Broadcasting & Cable*, p. 3.

McClellan, Steve. (1997b June 23). "Talk Up, Magazines Down." *Broadcasting & Cable*, p. 11.

McLaughlin, Lisa. (1993). "Chastity Criminals in the Age of Electronic Reproduction: Re-viewing Talk Television and the Public Sphere." *Journal of Communication Inquiry*, 17.1, 41–55.

McQuail, Denis. (1994). *Mass Communication Theory: An Introduction.* Third edition. London: Sage.

McRobbie, Angela. (1991). *Feminism and Youth Culture.* London: Macmillan.

Meehan, Eileen. (1994). "Conceptualizing Culture as Commodity," In *Television: The Critical View* Ed. Horace Newcomb. Fifth edition. New York: Oxford University Press.

Meehan, Eileen. (1991). "'Holy Commodity Fetish, Batman!': The Political Economy of a Commercial Intertext." In *The Many Lives of the Batman: Critical Approaches to a Superhero and His Media.* Eds. R.E. Pearson & W. Uricchio. NY: Routledge, pp. 47–65.

Messerschmidt, James. (1995). "Feminism, Criminology and the Rise of the Female Sex 'Delinquent,' 1880–1930." In *Gender, Crime and Feminism.* Ed. Ngaire Naffine. Brookfield, VT: Dartmouth, pp. 51–72.

Meyrowitz, J. (1985). *No Sense of Place.* New York: Oxford University Press.

Millner, Denene. (2000 December 12). "Iyanla: The New Talk of the Town." *Arts and Lifestyle: Television.*

Moore, Frazier. (2001 August 12). "Iyanla Brings Positive Approach to TV Talk Scene." *AP Online.*

Mosco, Vincent (1996). *The Political Economy of Communication: Rethinking and Renewal.* London: Sage.

Mukerji, Chandra and Michael Schudson. (Eds.). (1991). *Rethinking Popular Culture: Contemporary Perspectives in Cultural Studies.* CA: University of California Press.

Munson, Wayne. (1993). *All Talk: The Talkshow in Media Culture.* Philadelphia: Temple University Press.

Murdock, Graham. (1982). "Large Corporations and the Control of the Communications Industries." In *Culture, Society, and the Media.* Eds. Gurevitch, Michael et al. London: Methuen, pp 118–149.

Murdock, Graham and Peter Golding. (1974). "For a Political Economy of Mass Communication." In *The Socialist Register 1973* Eds. Milliband, R. and J. Saville. London: Merlin Press, pp. 205–234.

Murry, Velma McBride. (1996). "Inner-City Girls of Color: Unmarried, Sexually Active Nonmothers." In *Urban Girls: Resisting Stereotypes, Creating Identities*. Ed. Bonnie J. Ross Leadbetter and Niobe Way. New York: New York University Press, pp. 272–290.

The Negro Family: The Case for National Action. Office of Policy Planning and Research, U.S. Department of Labor. http://www. dol.gov/asp/programs/history/webid-moynihan.htm. Accessed 4–15–03.

Nelkin, Dorothy and M. Susan Lindee. (1995). *The DNA Mystique: The Gene as a Cultural Icon*. New York: Freeman.

Newcomb, Horace. (Ed.). (1994). *Television: The Critical View*. Fifth edition. New York: Oxford University Press.

Newfield, Christopher and Avery F. Gordon. (1996). "Multiculturalism's Unfinished Business." In *Mapping Multiculturalism*. Eds. Avery F. Gordon and Christopher Newfield. Minneapolis: University of Minnesota Press, pp. 76–115.

"Paternity Questions. (2002 May 6). "Are the Rumors True?" *Montel*. CBS. WUSA, Washington, DC.

Peck, Janice. (1994). "Talk About Racism: Framing a Popular Discourse of Race on Oprah Winfrey." *Cultural Critique*, 27, (Spring): 89–126.

Oprah's Habitat for Humanity Project. (1997). www.oprah.com. March.

Plummer, Ken. (1995). *Telling Sexual Stories: Power, Change and Social Worlds*. New York: Routledge.

Popenoe, David. "Life Without Father." In *Lost Fathers: The Politics of Fatherlessness in America*. Ed. Cynthia R. Daniels. New York: St. Martin's Press, 1998, 33–49.

Priest, Patricia Joyner. (1995). *Public Intimacies: Talk Show Participants and Tell-All TV*. Cresskill, NJ: Hampton Press, Inc.

Quail, Christine. (2003). *The Political Economy of Multiutilities*. Doctoral dissertation, University of Oregon.

Roach, Colleen. (1997). "Cultural Imperialism and Resistance in Media Theory and Literary Theory." *Media, Culture & Society*, 5.19: 47–66.

Rothman, Barbara Katz. (1994). "Beyond Mothers and Fathers: Ideology in a Patriarchal Society." In *Mothering: Ideology, Experience, and Agency*. Eds. Evelyn Nakano Glenn et al. New York: Routledge, pp. 139–157.

Russo, Mary. (1994). *The Female Grotesque: Risk, Excess and Modernity*. New York: Routledge.

Sachs, Andrea. (2001). "Can They De-Springerize Talk? These hosts say viewers want a new alternative to "screamer" shows. Oprah, meet Iyanla and Ananda." *Time*, 6 August, 56.

Saltzman, Joe. (Nov 1996). "Why Ordinary Americans Like Daytime Talk Shows," *USA Today: The Magazine of the American Scene*. 5.125.2618: 63.

Sandler, Adam. (1996 December 23–1997 January 5). "Warblers Warm Up at Oprah House," *Variety*, pp. 1, 58.

Schiller, Herbert. (1989). "Thinking About Media Power: Who Holds It? A Changing View." In *Culture Inc*. New York: Oxford University Press.

Schiller, Herbert. (1991). "Not Yet the Post-Imperialist Era." *Critical Studies in Mass Communication* 8, pp. 13–28.

Scott, Gini Graham. (1996). *Can We Talk? The Power and Influence of Talk Shows*. New York: Insight.

Seiter, Ellen. (1981). "The Promise of Melodrama: Recent Women's Films and Soap Operas." Doctoral dissertation. Northwestern University.

"Send My Wild Teen to Boot Camp." Air date 8–03–00.

Sennett, R. (1977). *The Fall Of Public Man*. New York: Alfred A. Knoff.

Shattuc, M. Jane. (1997). *The Talking Cure: TV Talk Shows and Women*. New York: Routledge.

Smythe, Dallas. (2001). "On the Audience Commodity and Its Work." In *Media and Cultural Studies: Key Works* Eds. M. G. Durham and D. Kellner. London: Blackwell.

Solinger, Ricki. (1999). "Dependency and Choice: The Two Faces of Eve." *Whose Welfare?* Ed. Gwendolyn Mink. Ithaca, NY: Cornell University Press.

Solinger, Ricki. (2000). *Wake Up Little Suzie*. New York: Routledge.

Stallybrass, P. and A. White. (1986). *The Poetics and Politics of Transgression*. London: Methuen.

Stam, R. (1982). "On the Carnivalesque." *Wedge*, 1: 47–55.

Stanley, Alessandra. (1997). "Mother Russia Meets Dr. Kinsey on TV Talk Show." *New York Times*, November 14. [on AOL].

Starbucks Raises $185,000 to Support Literacy! (1997 September 23). Oprah Online. www.oprah.com

Starbucks and Oprah Fight Illiteracy. (1997 May 28). Oprah Online. www.oprah.com.

Steenland, Sally. (1994). "Those Daytime Talk Shows." *Television Quarterly*, 43.4: 5–12.

Steinberg, Shirley and Joe Kincheloe. (Eds.) (2004). *Kinderculture: The Corporate Construction of Childhood*. Second Edition. Boulder: Westview.

Stern, Christopher (1995 October 30). "Backlash Against TV Talk Shows." *Broadcasting & Cable*, p.18.

Sturken, Marita and Lisa Cartwright. (2001). *Practices of Looking: An Introduction to Visual Culture*. New York: Oxford University Press.

Thompson, Kenneth. (1998). *Moral Panics*. New York: Routledge.

Tolman, Deborah. (1996). "Adolescent Girls' Sexuality: Debunking the Myth of the Urban Girl." In *Urban Girls: Resisting Stereotypes, Creating Identities*. Eds. Bonnie J. Ross Leadbetter and Niobe Way. New York: New York University Press, pp. 255–271.

Tucker, Robert C. (Ed.) (1978). *The Marx-Engels Reader*. Second edition. New York: W.W. Norton & Company.

Turner, Graeme. (1996). *British Cultural Studies: An Introduction*. Second edition. London: Routledge.

Twitchell, James. (1992). *Carnival Culture: The Trashing of Taste in America*. New York: Columbia University Press.

Van Dijck, Jose. (1998). *Imagenation: Popular Images of Genetics*. New York: New York University Press.

Vedder, Clyde and Dora Sommerville. (1975). *The Delinquent Girl*. Springfield, IL: Charles C. Thomas Publishing.

Wagner, R. Helmut. (Ed.). (1970). *Alfred Schutz: On Phenomenology and Social Relations. Selected Writings*. Chicago: The University of Chicago Press.

Wasko, Janet. (1995). *Hollywood in the Information Age*. Austin: University of Texas Press.

Wasko, Janet. (2001). *Understanding Disney: The Manufacture of Fantasy*. Cambridge: Polity Press.

Wasko, Janet and Vincent Mosco. (Eds.). (1992). *Democratic Communication in an Information Age*. Toronto: Garamond; and Norwood, NJ: Ablex.

White, Mimi. (1992). *Tele-Advising: Therapeutic Discourse in American Television*. London: University of North Carolina Press.

"The Wildest, Most Memorable Maury Guests" Maury. Fox. WTTG, Washington, D.C. 6 May 2002.

Williams, Montel. http://www.montelshow.com/montel/ashesees. htm, accessed 4–30–02.

Williams, Raymond. (1958). *Culture and Society*. London: Chatto and Windus.

Williams, Raymond. (1961). *The Long Revolution*. London: Chatto and Windus.

"Who Is the Father of My Baby" *Jenny Jones*. Fox. WTTG, Washington, D.C. 30 July 2002.

Wolf, Naomi. (1991) *The Cult of Beauty: How Images of Beauty Are Used Against Women*. New York: Doubleday.

Woodhull, Nancy J. and Robert W. Snyder. (1998). *Media Mergers*. New Brunswick/London: Transaction Publishers.

Working for Independence. "Promoting Child Well-Being and Healthy Marriages." http://www.whitehouse.gov/news/releases /2002/02/welfare-book-05.html, accessed 4–15–03.

www.cbs.com/specials/dr_phil/

http://disney.go.com/home/today/index.html

http://www.drphil.com/about/about_landing.jhtml

www.jennyjones.warnerbros.com, Weekly Show Schedule.

www.oprah.com

www.sallyjr.com/sally4/frm_bootcamps.html

www.spe.sony.com/tv/shows/ricki, Today's Show and Coming Soon.

www.studiosusa.com/maury, Episode Guide

www.tvtalkshows.com.

Studies in the Postmodern Theory of Education

General Editors
Joe L. Kincheloe & Shirley R. Steinberg

Counterpoints publishes the most compelling and imaginative books being written in education today. Grounded on the theoretical advances in criticalism, feminism, and postmodernism in the last two decades of the twentieth century, Counterpoints engages the meaning of these innovations in various forms of educational expression. Committed to the proposition that theoretical literature should be accessible to a variety of audiences, the series insists that its authors avoid esoteric and jargonistic languages that transform educational scholarship into an elite discourse for the initiated. Scholarly work matters only to the degree it affects consciousness and practice at multiple sites. Counterpoints' editorial policy is based on these principles and the ability of scholars to break new ground, to open new conversations, to go where educators have never gone before.

For additional information about this series or for the submission of manuscripts, please contact:

Joe L. Kincheloe & Shirley R. Steinberg
c/o Peter Lang Publishing, Inc.
275 Seventh Avenue, 28th floor
New York, New York 10001

To order other books in this series, please contact our Customer Service Department:

(800) 770-LANG (within the U.S.)
(212) 647-7706 (outside the U.S.)
(212) 647-7707 FAX

Or browse online by series:

www.peterlangusa.com